Summer Counts!

2nd Edition

Grade 8→9

Thematic Reading, Language Arts, and Math Skills

AF860390

Options™

ISBN 978-1-60161-933-4

OT108

Cover Image: Ian Keltie/Shannon Associates, LLC

Triumph Learning® 136 Madison Avenue, 7th Floor, New York, NY 10016

Printed in the United States of America.

10 9 8 7 6 5 4 3 2 1

Dear Parent,

Summer is a time for relaxing and having fun. It can also be a time for learning. *Summer Counts!* can help improve your child's understanding of important skills learned in the past school year while preparing him or her for the year ahead.

Summer Counts! provides grade-appropriate practice in subjects such as reading, language arts, vocabulary, and math. The ten theme-related chapters include activities and puzzles to motivate your child throughout the summer.

When working through the book, encourage your child to share his or her learning with you. You may want to tear out the answer key at the back of the book and use it to check your child's progress. With *Summer Counts!* your child will discover that learning happens anytime—even in the summer!

Apreciados padre,

El verano es una época para descansar y divertirse. También puede ser una época para aprender. *Summer Counts!* puede ayudar a que su hijo(a) mejore las destrezas importantes que aprendió el pasado año escolar al mismo tiempo que lo(a) prepara para el año que se aproxima.

Summer Counts! provee la práctica apropiada para cada grado en las asignaturas como la lectura, las artes del lenguaje y las matemáticas. Los diez capítulos temáticos incluyen actividades y rompecabezas que motivarán a su hijo(a) durante el verano.

Cuando trabaje con el libro, anime a su hijo(a) a que comparta lo que ha aprendido con Ud. Si Ud. desea puede desprender la página de las respuestas que aparece en la parte trasera del libro. Puede usar la misma para revisar el progreso de su hijo(a). ¡Con *Summer Counts!* su hijo(a) descubrirá que el aprendizaje puede ocurrir en cualquier momento—inclusive en el verano!

Table of Contents

CHAPTER 1

The Snake and the Buzzards

A snake once met two buzzards sitting on a rock and said to them, "How fortunate you two are. You can travel through the air, while I always go with my stomach to the ground. How wonderful it must be to sail through the blue sky, and how unlucky I am always crawling on the ground."

The two buzzards listened and felt sorry for him. They were kind birds who never hurt another animal. They always tried to help animals and mankind by cleaning up the countryside. The older bird said, "Things are not always as pretty as they look. Sometimes it becomes tiresome living in the air. We are not welcome on the ground, you know. With you it is different. Everyone is afraid of you."

"Ah, yes," sighed Snake. "But at heart, I am like you two. By nature I am really a bird. That is what I am. Oh, if I could only fly once, it would be so wonderful."

"We buzzards are not very strong—mostly feathers. Feel. We cannot carry you. But there are more ways of doing things than what's been tried. I know how we can take you for a ride through the blue sky. We will both carry you; that is what we will do." The two birds then went to the nearby desert and brought back a dried-up yucca stalk. The stalk was about six feet long. It was lightweight, but strong.

Each bird took an end of the stick in his mouth and told the snake to follow them up a small hill. "Take a tight hold here in the middle of the stick with your mouth. We will each take hold of an end and fly with it in our beaks. Hold on tight." The birds went running and flapping their wings down the hill and shouting to each other, "Both together, now." At the bottom of the hill, they rose into the air with the snake holding fast to the stick with his long fangs.

Soon a great crowd of birds gathered behind them. They all wondered what this strange group of flyers might be. Eagle flew close and began to say things to make Snake angry. "Flap your wings, Brother Snake," he said. "Do I see feathers sprouting from your tail?"

Then Dove came and perched on the yucca stalk. "Don't answer him, Snake. He is trying to make you open your mouth to talk so you will lose your hold and fall to the ground." Seeing Dove so close to him made Snake forget for a moment that he was not a bird. He opened his jaws to make a grab for Dove. As soon as he did, he dropped from the stick, fell through the air, and landed on a prickly pear bush on the ground. The fall knocked the breath out of him and countless cactus thorns into him. He crawled back to his den in the rocks. In the days that followed, every time he pulled out a thorn he cared a little less about flying.

The Snake and the Buzzards

Directions **Using what you have just read, answer the questions.**

1. Why did Snake think he was unlucky?

2. How are the buzzards helpful to animals and mankind?

3. Why do you think Eagle wanted to make Snake angry?

4. What happened after Dove perched on the stick?

5. Why do you think Snake grabbed for Dove?

6. How are the buzzards and Dove alike?

7. Do you think Snake will ever fly again?

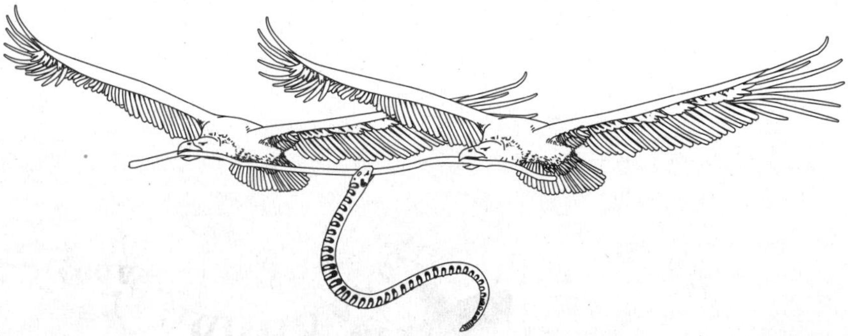

Adverbs: Comparative and Superlative

REMEMBER

An **adverb** describes an action verb, an adjective, or another adverb. It can tell *when, where, how*, or *to what extent* an action happens.
A **comparative adverb** compares two actions.
A **superlative adverb** compares more than two actions.

EXAMPLES

Adverb: Owls *sometimes* nest in prairie dog burrows. (*when*)
Comparative adverb: Do foxes run *faster* than giraffes? (*how*)
Superlative adverb: What animal runs *fastest* of all? (*to what extent*)

Adverb Antics

Directions **Draw a line under each adverb. Then write if the adverb is comparative or superlative on the line.**

1. Can you accurately identify the birds in your area? ________________

2. Stick insects can hold their twig-shaped bodies motionless for hours. ________________

3. Which animal swims most gracefully of all? ________________

4. The walrus hauled himself clumsily over the snow. ________________

5. Bighorn sheep climb easily over narrow mountain ledges. ________________

6. Some octopuses change color more frequently than others. ________________

7. A termite mound was quickly built under the wood deck. ________________

8. The field mouse perches delicately on the grass stalk. ________________

9. Piranhas attack more ferociously when in groups. ________________

10. The grizzly bear snatched hungrily at the jumping salmon. ________________

Animal World

Directions **Choose words from the box below that match each definition. Write them on the lines.**

aggressive	anatomy	buoyant	defenseless
edible	overpower	predators	prey

1. Exposed or unprotected ____________________

2. Hostile ____________________

3. Safe to eat ____________________

4. A victim, the hunted ____________________

5. Having the ability to float ____________________

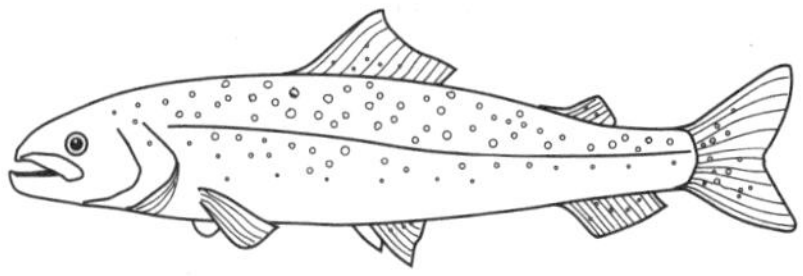

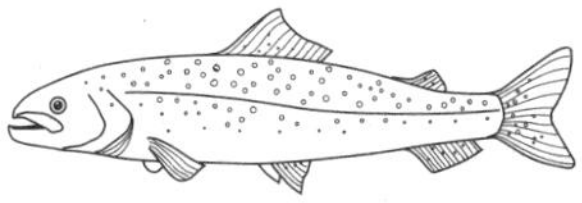

Hammerhead Sharks

Directions **Read the article. Use words from the box above to fill in the blanks.**

Hammerhead sharks hunt other animals and eat decaying matter, making them both **(6)** ____________________ and scavengers. They are called hammerheads because of their distinct **(7)** ____________________ and the strange shape of their head. It is flat and shaped like the head of a hammer. Hammerheads will feed on just about anything, but they typically eat a variety of fish including skates, sting rays, and smaller hammerheads. Larger hammerheads have even been known to become **(8)** ____________________ toward humans. This shark has incredible speed and strength to **(9)** ____________________ its prey. Being unable to stay afloat, or **(10)** ____________________, hammerheads must swim constantly to keep from sinking.

Prime and Composite Numbers

REMEMBER

A **factor** of a number divides the given number evenly. When you divide a number by one of its factors, there is no remainder. A **composite number** has more than two factors. The number 1 is neither prime nor composite.

EXAMPLES

The number 4 has 3 factors—1, 2, and 4.
The number 4 is a composite number.

$$4 \div \mathbf{1} = 4$$

$$4 \div \mathbf{2} = 2$$

$$4 \div \mathbf{4} = 1$$

The number 17 has two factors—1 and 17.
The number 17 is a prime number.

$$17 \div \mathbf{1} = 17$$

$$17 \div \mathbf{17} = 1$$

Fierce Factors

Directions Write the factors of each number. Then circle whether the number is **prime** or **composite** on the lines.

1. 5 ____________________ **prime** **composite**

2. 8 ____________________ **prime** **composite**

3. 17 ____________________ **prime** **composite**

4. 22 ____________________ **prime** **composite**

5. 36 ____________________ **prime** **composite**

6. 51 ____________________ **prime** **composite**

7. 73 ____________________ **prime** **composite**

8. 81 ____________________ **prime** **composite**

9. 127 ____________________ **prime** **composite**

10. 87 ____________________ **prime** **composite**

The Zoology Test

Directions **Mr. Garcia gave his students a test on the zoology unit they just completed. The test contained 40 questions about different animals. Mr. Garcia listed the 25 scores of his students. Below are the scores. Use the information to complete the page.**

34, 25, 38, 36, 33, 31, 36, 37, 38, 29, 28, 22
38, 30, 29, 40, 39, 35, 20, 26, 28, 38, 35, 30, 36

1. What is the mean of the student scores? ________

2. What is the median of the student scores? ________

3. What is the mode of the student scores? ________

4. How many students scored:

a. above 80%? ________

b. from 60% to 80%? ________

c. below 60%? ________

5. Complete the line plot below to show the distribution of the scores.

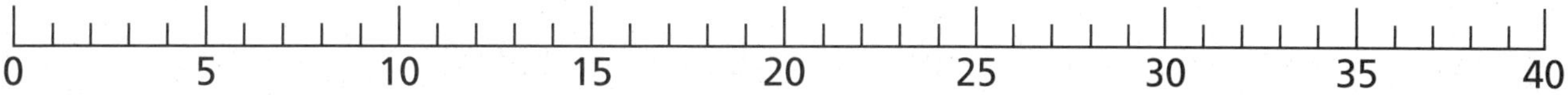

Bird Rhymes

Directions **Use the clues to solve the riddles. Hint: Think about words that rhyme with those in dark print.**

EXAMPLE: What do you call group of **birds**?

A bird herd

1. What do you call a **week**-long bird convention?

   A ____________ week

2. What do you call a search for a new bird **nest**?

   A nest ____________

3. What do you call good flying **weather** for birds?

   The ____________ weather

4. What do you call a group of eggs that are ready to **hatch**?

   A hatch ____________

5. What do you call a bird's **tail** that has light-colored feathers?

   A ____________ tail

6. What do you call the **height** a bird flies at?

   The ____________ height

Mean, Median, and Mode

Directions Eighteen students raised money to donate to an aviary that helps sick and injured birds. The list below shows the number of dollars each student raised.

84, 112, 147, 36, 99, 221, 140, 150, 85, 122, 157, 62, 255, 101, 147, 88, 96, 148

1. What is the mean of this data set?

2. What is the median of this data set?

3. What is the mode of this data set?

4. How many students raised

a. more than $150?

b. between $100–$150?

c. less than $100?

CHAPTER 2

The *Heart* of the Matter

Long ago the heart was thought to be the origin of love and courage. Now we know that the brain controls these feelings, but the heart is still the symbol of love. Because the heart is one of the most powerful organs in the body, it should also be the symbol of strength. The world's best engineers could not have designed a more efficient pumping machine!

The heart is a hardworking muscular pump about the size of a clenched fist. It lies near the midpoint of the upper chest. The muscles of the heart contract and relax, sending blood to all parts of the body. This pattern of contraction and relaxation happens about 70 times per minute.

Each heartbeat sends a high-pressure surge of blood into the arteries, which are blood vessels that have elastic walls. You can feel these surges of pressure at various points on the body, such as the inside of your wrist and the side of your neck. These pulse points make it easy to determine how hard and how quickly the heart is beating.

The heart is actually two pumps that work side by side. The right and left sides are separated by a partition, or wall, of muscle. Each pump has an upper chamber called the atrium (AY-tree-um). Below each atrium is a larger chamber called the ventricle (VEN-trih-kul). This is the source of the heart's main pumping power. The muscle in each ventricle squeezes hard during each heartbeat, forcing blood out. One-way valves prevent blood from flowing back into each chamber.

The movement of blood from the heart to the lungs and back to the heart is called pulmonary circulation. Systemic circulation moves blood to all of the body tissues except the lungs.

The heart is a versatile organ. It responds to nerve signals from the brain and hormones in the blood. The heart alters its pumping speed and forces it to match the body's needs. Different situations and activities can affect the rate at which the heart pumps blood. Sleep makes the heart slow down, while exercise or stress increase the heart rate. Unlike other muscles in the body, the heart muscle never stops. It works both day and night.

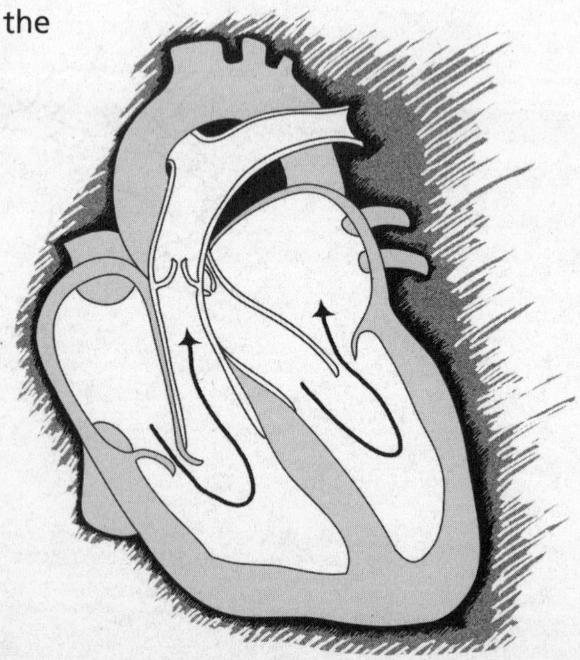

The Heart of the Matter

Directions **Using what you have just read, answer the questions.**

1. What part of the body controls feelings?

____

2. What makes the heart beat?

____

____

3. Where does the blood go when it leaves the heart?

____

4. Why does this article say the heart is like two pumps?

____

____

5. How is pulmonary circulation the same as systemic circulation? How are they different?

____

____

6. How does exercise change the rate at which the heart pumps blood?

____

____

7. Why did the author write this article?

____

____

____

Gerunds and Gerund Phrases

REMEMBER

A **gerund** is a verb form ending in *-ing* that is used as a noun. A gerund acts as a subject, a direct object, an indirect object, an object of a preposition, a predicate noun, and an appositive.

A **gerund phrase** is a group of words that includes a gerund and one or more other words that work together as a noun.

EXAMPLES

Gerund: *Staying* healthy involves more than just physical health.

Gerund phrase: *Staying in shape* takes only a few hours a week.

Gerund Exercises

Directions Draw a line under each **gerund** or **gerund phrase** in the sentences below.

1. Recognizing healthy choices is the first step.
2. Looking at your choices helps you decide if you need to make changes.
3. Good health also means feeling good about yourself.
4. Accepting yourself includes liking who you are.
5. Exercising is one choice that affects your health.
6. Social health involves getting along with others.
7. Pamela gave a speech about communicating effectively.
8. Dealing with stress in a positive way is important for good health.
9. Practicing good health habits is the key to wellness.
10. Jim hopes to accomplish his goal, running three miles every day.

Staying Fit

Directions **Choose words from the box below that match each definition. Write them on the lines.**

adolescence	calories	confidence	encourage
nutritious	prevent	healthy	diet

1. To avoid or keep from occurring ______________________
2. The energy value of food ______________________
3. Teenage years are called this ______________________
4. Self-belief, assurance ______________________
5. To support someone or something ______________________

A Healthy Attitude

Directions **Read the article. Use words from the box above to fill in the blanks.**

The best way to stay **(6)** ______________________ is to keep active and exercise. A balanced **(7)** ______________________ of fruits and vegetables helps keep our bodies performing at their best. These **(8)** ______________________ foods give us the energy we need to keep going throughout the day. You should also be sure to eat enough **(9)** ______________________ so you're not hungry. Eating right and exercising can also help boost self- **(10)** ______________________. Sports aren't the only way to stay active. Just put on a pair of sneakers and start moving!

Order of Operations

REMEMBER

When you evaluate mathematical expressions, you should follow the order of operations.

1. Do operations within the parentheses or square root symbol first.
2. Do all exponents before other operations.
3. Multiply and divide in order from left to right.
4. Add and subtract in order from left to right.

EXAMPLE

$6 \times (7 - 5) - 100 \div 25$	$(7 - 5) = 2$
$\mathbf{6 \times 2 - 100 \div 25}$	$\mathbf{6 \times 2 = 12,\ 100 \div 25 = 4}$
$\mathbf{12 - 4}$	$\mathbf{12 - 4}$
8	**8**

Directions **Find the value of each expression. Show your work.**

1. $24 - 15 \div 3 + 2 \times 4$

2. $\sqrt{64} + (4 - 1)^2 - 12 \div 3$

3. $(1 + 1)^3 + 4 \times 5 - \sqrt{16}$

4. $15 - 6\,(5 + 3) \div 4$

A Friendly Probability

Directions **Susan, Lia, and Monique have tickets to a play. They have three seats together in row E. Their ticket numbers are E9, E10, and E11. Use this information to complete the page.**

1. What is the probability that Lia will sit in one of the outside seats (seat E9 or seat E11)? Express the probability as a fraction. ________

2. Complete the list of the different seating arrangements that are possible for the three girls.

E9	E10	E11
Susan	Lia	Monique

3. How many different seating arrangements are possible? ____________

4. What is the probability that Lia and Susan will sit next to each other? Express the probability as a fraction. ____________

5. A drawing will be held at the play and 3 door prizes will be given out. Each person who attends has a chance of winning. If 100 people attend the play, what is the probability that one of the three girls will win the first prize? Explain your answer.

____

____

Feeling Stressed

Directions **Read each definition. Unscramble the letters to find the words. Write the words on the lines. Then use the numbered letters to solve the riddle.**

1. reabht You can take a deep one of these to release some tension

___ ___ ___ ___ ___ ___
(letter 5 = 1)

2. chadeahe A throbbing pain in the head sometimes caused by stress

___ ___ ___ ___ ___ ___ ___ ___
(letter 3 = 2)

3. xrlae To calm down when stressed

___ ___ ___ ___ ___
(letter 3 = 3)

4. remokowh Work given at school to be done that night

___ ___ ___ ___ ___ ___ ___ ___
(letter 8 = 4)

5. odlob The heart pumps this through the body

___ ___ ___ ___ ___
(letter 3 = 5)

6. sestsr Anxiety or worry that can be unhealthy

___ ___ ___ ___ ___ ___
(letter 1 = 6)

7. stimeono Anger, love, and frustration are different types of these

___ ___ ___ ___ ___ ___ ___ ___
(letter 2 = 7)

8. treha This may beat faster in stressful situations

___ ___ ___ ___ ___
(letter 2 = 8)

9. npal To prepare for what is ahead

___ ___ ___ ___
(letter 4 = 9)

NOW SOLVE THE RIDDLE:

What is the best way to reduce stress?

___	___	___	___		___	___		___	___	___	___	___	___	___ !
1	2	3	4		1	5		6	5	7	8	5	9	8

Above and Below

REMEMBER

When an expression has operations above and below a division bar, treat these operations as if they were in parentheses.

EXAMPLE

$$\frac{14 + 4}{8 - 2} = \frac{(14 + 4)}{(8 - 2)} = \frac{18}{6} = 3$$

Directions **Find the value of each expression. Show your work.**

1. $\frac{13 + 8}{3 + 4}$

2. $\frac{56 \div 8}{20 - 13}$

3. $\frac{25 + 8}{3 + 72 \div 9}$

4. $\frac{20 + (100 \div 25)}{22 - 8 \times 2}$

5. $\frac{(2 + 3)^3 + 3}{19 + 2 \times 4 + 5}$

CHAPTER 3

Vertically Challenged

Imagine yourself clinging to the side of a sheer rock wall by your fingers and toes, two thousand feet above the ground. Does the thought terrify you? If it sounds more exciting than scary, then rock climbing may be the sport for you!

There are actually several types of rock climbing:

- *traditional*—climbers place and remove gear as they go
- *sport climbing*—on walls with fixed gear
- *bouldering*—climbing big rocks on, or close to, the ground
- *ice climbing*—climbing frozen waterfalls
- *top roping*—using an anchor at the top of the climb

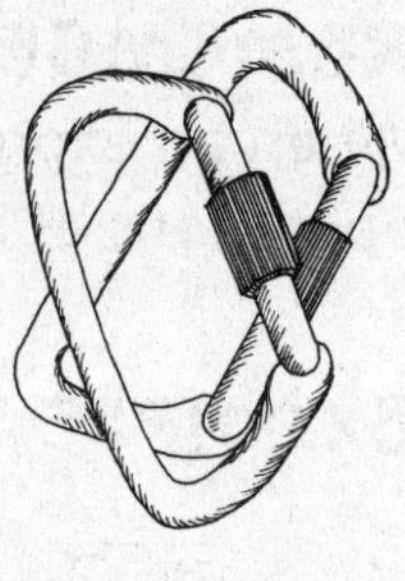

Most beginners start with bouldering. This type of climbing doesn't require a safety rope, because climbers don't fall far enough to get injured. Falling, by the way, is an inevitable part of the sport of rock climbing. Beginning climbers look for boulders with easily angled ridges and rocks that have cracks and ledges of all sizes.

Basic rock-climbing gear includes special shoes with sticky soles that adhere to footholds; a helmet made of a durable, but lightweight plastic; and rope, which acts as a safety net to stop a fall. These ropes need regular care and replacement every few months or years, depending on how often you climb.

Climbers also use a climbing harness, a belay device, and carabiners. The harness supports the climber. A belay device is a hand-operated brake that slows the rope. As you climb a rock, a second climber on the ground, called the belayer, takes up slack on the rope. If you fall, the tightened rope breaks your fall. Carabiners are C-shaped aluminum rings used to clip two pieces of gear together. They are also used to clip gear to the rope.

The United States has an outstanding array of climbing areas. One of the most popular rock faces for expert climbers is the 3,000-foot El Capitan ("El Cap"), in Yosemite National Park, California. The sheer, granite wall has several climbing routes of extreme difficulty.

Rock climbing is a dangerous, but exciting sport that demands both brains and brawn. It requires confidence and calmness under pressure, as well as stamina, good balance, and agility. If you think you'd like to try rock climbing, the best place to start is on a wall at an indoor climbing gym. You can learn the basics from instructors there. It will give the phrase "climbing the walls" a whole new meaning!

Vertically Challenged

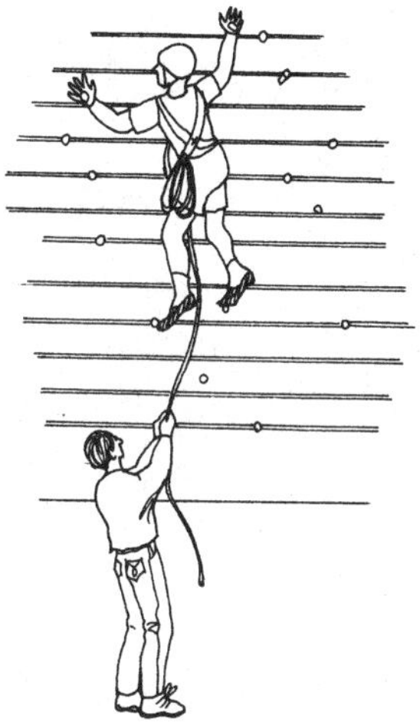

Directions Using what you have just read, answer the questions.

1. What is the main idea of this article?

2. How is traditional rock climbing different from sport climbing?

3. Why do you think most beginning climbers start with bouldering?

4. Why do you think the ropes used in rock climbing would need regular care?

5. Would you rather be the climber or the belayer? Explain.

6. What does *brawn* mean?

7. How does the author feel about rock climbing? How do you know?

Participles: Past and Present

REMEMBER

A **participle** is a verb form used as an adjective that modifies a noun or pronoun. **Present participles** end in *-ing*. **Past participles** often end in *-ed*, but they can also have irregular endings such as *-en*, *-t*, or *-n*.
A **participle phrase** is a group of words that includes a participle used as an adjective and other words that complete its meaning.

EXAMPLES

Present participle: Hanna laced her *running* shoes.
Past participle: She had a *worried* expression on her face.
Participle phrase: *Sitting quietly by herself*, Hanna thought about the race.

First-Place Participles

Directions **Draw a line under each participle used as an adjective.**

1. Hanna could tell that it would be a scorching day.
2. The excited participants stretched or jogged in place.
3. A voice on the loudspeaker announced the scheduled events.
4. Hanna looked at her worn shoes and smiled.
5. These shoes soothed her aching feet.
6. She had decided to wear them in the upcoming meet.
7. The waiting runners gathered on the grassy field.
8. "I'll need water for my parched throat!" cried Sylvia.
9. Hanna shared her bottled water with Sylvia.
10. Then she tried to calm her pounding heart.

Fast Feet

Directions **Choose words from the box below that match each definition. Write them on the lines.**

agile	circulation	dehydrated	energetic	gear
injury	jogger	muscles	surface	endurance

1. Dried out or lacking water ____________________
2. Active or lively ____________________
3. Nimble or quick on your feet ____________________
4. Equipment used in a sport ____________________
5. Stamina or ability to continue ____________________

Finish the Article

Directions **Read the article. Use words from the box above to fill in the blanks.**

Millions of people jog to stay fit. Jogging builds strong **(6)** ____________________ and burns calories, so it can help to control your weight. It also improves breathing, blood **(7)** ____________________, and decreases the risk of heart disease. However, jogging can cause problems if not done properly. A **(8)** ____________________ should run on a soft **(9)** ____________________ and build up his or her distance gradually to avoid **(10)** ____________________. Joggers should also drink plenty of water so they don't get **(11)** ____________________. To start a jogging routine, you can begin by walking, then increase your speed as you become more **(12)** ____________________ on your feet!

Finding Proportions

REMEMBER

A **proportion** is an equation that shows that two ratios are eqivalent. In a proportion, the cross products are equal.

EXAMPLE

$$\frac{2}{3} = \frac{8}{12}$$

$3 \times 8 = 2 \times 12$ ← cross products

$$\mathbf{24 = 24}$$

True Proportions

Directions Write **yes** if the equation shows a proportion. Write **no** if it does not show a proportion.

1. $\frac{3}{4} = \frac{11}{15}$ ________

2. $\frac{2}{3} = \frac{24}{36}$ ________

3. $\frac{2}{5} = \frac{11}{25}$ ________

4. $\frac{5}{9} = \frac{35}{63}$ ________

Directions Write the missing number in each of the following proportions.

5. $\frac{2}{3} = \frac{10}{\_\_}$

6. $\frac{3}{\_\_} = \frac{36}{48}$

7. $\frac{\_\_}{8} = \frac{54}{72}$

8. $\frac{7}{12} = \frac{\_\_}{60}$

9. $\frac{12}{72} = \frac{2}{\_\_}$

10. $\frac{8}{2} = \frac{\_\_}{11}$

Graphing Results

Directions **Ahmad volunteers for a physiological profile. His oxygen-use rate and calorie-use rate are shown in the graphs below. Study the graphs and then complete the page.**

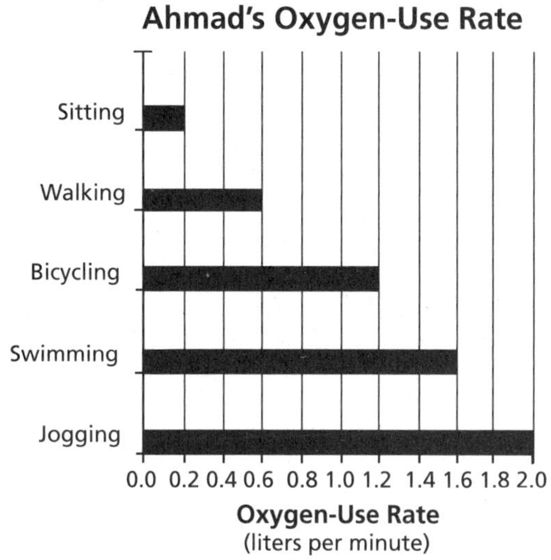

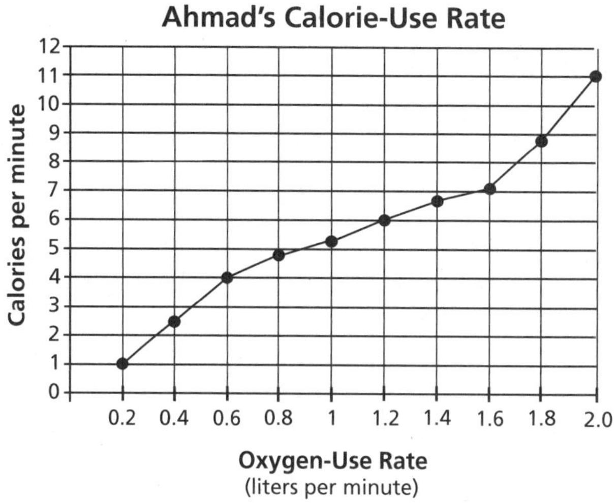

1. What is Ahmad's oxygen-use rate for each activity? Write your answers in Column A in the table.

2. How many more liters of oxygen does Ahmad use each minute jogging than he does sitting?

____

3. What is Ahmad's calorie-use rate for each activity? Write your answers in Column B.

4. About how many calories per minute does Ahmad use while he is walking?

____

swimming? _______________

Activity	Oxygen-Use Rate (A)	Calorie-Use Rate (B)
Sitting		
Walking		
Bicycling		
Swimming		
Jogging		

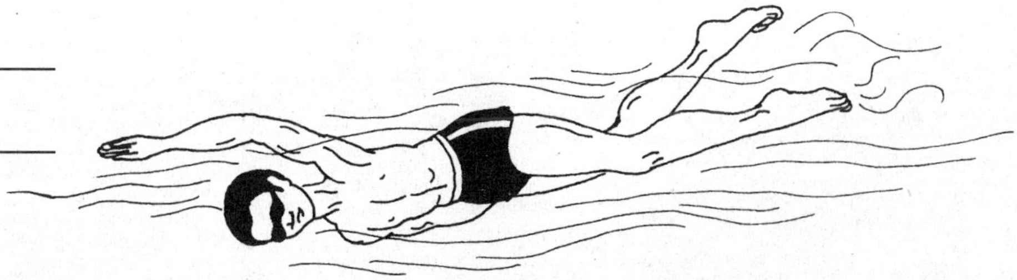

Picture-Perfect Sports

Directions Use the picture clues below to name the sport. Then write it on the lines.

EXAMPLE: 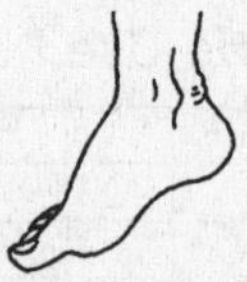 +  = football

1.  +  = ____________

2.  + cl +  + ming = ____________

3. 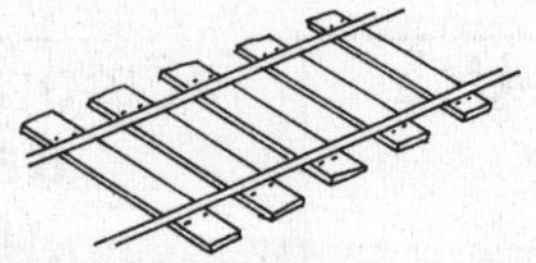 = ____________

4. h +  j + 

= ____________

5. s +  + ing = ____________

6. sk + 8 +  + 

= ____________

7. ch +  +  + ing = ____________

Julio's Ten-Kilometer Race

Gregory was on the football team, but he wanted to get into better shape. He told Julio he needed a partner to motivate him to run every afternoon. Julio wasn't particularly athletic, but he decided to give it a try.

Gregory and Julio enjoyed running so much, they decided to enter a ten-kilometer race. However, the day before the race, Gregory sprained his ankle playing football. Julio didn't feel confident, but he decided to run alone.

Julio ran the first kilometer easily, but then his legs started to feel heavy, and he felt out of breath. Another runner shouted, "Slow down! You'll make it!" Julio took the runner's advice. He didn't win, but he wasn't the last person to cross the finish line either! He heard Gregory shouting from the crowd, "Way to go, Julio!"

Directions **Circle the letter of the best answer.**

1. Why did Gregory ask Julio to run with him?

A. Gregory was afraid to run alone.

B. Gregory thought Julio needed to exercise.

C. Gregory thought a running partner would motivate him.

D. Gregory thought Julio would teach him how to run faster.

2. Which of the following is true about both Julio and Gregory?

A. They are both athletic.

B. They both enjoy running.

C. They both run in a ten-kilometer race.

D. They both play football.

3. What is the meaning of the word **confident** in the passage?

A. angry **B.** comfortable

C. upset **D.** confused

4. Which of the following words in the passage is an adverb?

A. partner **B.** motivate

C. athletic **D.** easily

CHAPTER 4

How It All Began

By Nora Evangeline

In school, I always filled my notebooks with sketches and written notes. It only seemed natural that the pages of the journal I started in Africa became filled with drawings of people and landscapes. I also drew in the margins of the weekly letters I sent to my father back in Atlanta. I saw so many new things while I was on a nursing assignment in Africa that I just had to put them down on paper in both words and pictures. I didn't know, however, that my father was printing the letters I sent him in a newspaper column called "Letters from Africa." After several months, he sent me a few of the clippings. I was pleased that he thought my letters were good enough to print, but I was embarrassed. Knowing others would see them encouraged me to be more careful with my words and pictures.

When the first letter from Bennett Wetherill at Fairlawn Press reached me in the Congo, I confess, I had never heard of him or his publishing company. He seemed like a nice man, though. He wanted to know if I had any "material" beside what was appearing in the newspaper. I had filled the bound notebook I was using as a journal and had just started another one. So, without much thought, I stuck my journal into an envelope and mailed it to him. When a book contract arrived in the mail eight weeks later, I was shocked. What really surprised me was the amount of money Fairlawn Press had enclosed with the contract! After a telephone conversation with Wetherill, the deal was concluded. I tucked the check away and began making plans to take a vacation in Europe on my way back to Atlanta.

Congo Journal was published while I was stopping over in Rome. All those museums filled with glorious art had further inspired me. I bought a box of oil paints and an easel. I wanted to try my hand at something more substantial than the ink and colored pencils I had been working with. I was staying at a small, residential hotel in the hills outside Rome when a note from Alan Burke at *The New Yorker* magazine reached me. He asked if I would travel to Israel to write and illustrate an article. "Why not try it?" I thought. The money they were offering for a few weeks' work would allow me to return to Rome for another four or six weeks when I was done! I arrived in Tel Aviv on June 5, 1967, at the beginning of the Six Day War. From there, I began writing my second book, *War and Peace in the Desert*.

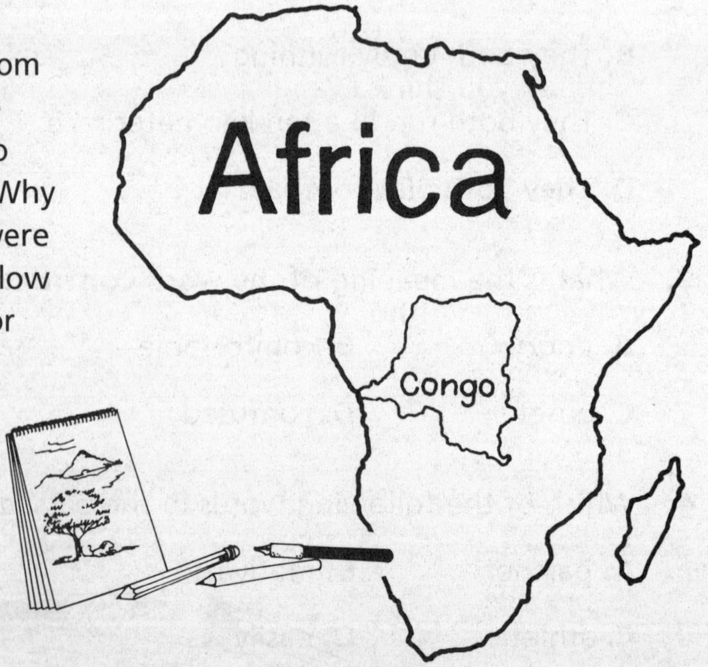

How It All Began

Directions **Using what you have just read, answer the questions.**

1. Why was Nora Evangeline in the Congo?

2. What is "Letters from Africa"?

3. How did Evangeline's letters to her father change when she found out other people were reading them?

4. How do you think that Bennett Wetherill found out about Evangeline's letters to her father?

5. Do you think Evangeline liked Rome? Why or why not?

6. What do you think Evangeline's new occupation will be?

7. Why did the author write this selection?

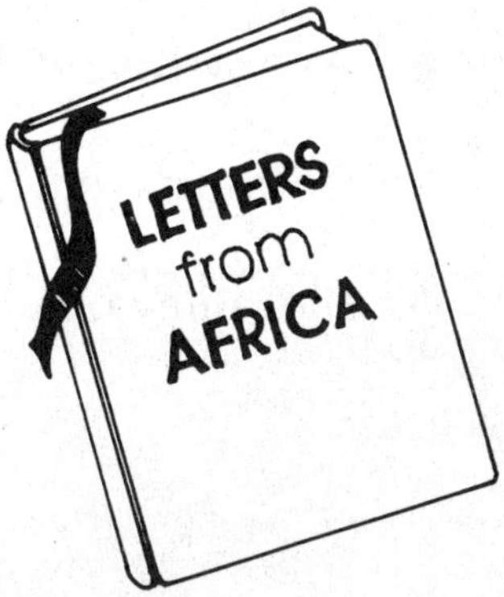

Infinitives

REMEMBER

An **infinitive** is a verb form that usually begins with *to* and can be used as a noun. An infinitive used as a noun can be the subject, predicate noun, or direct object in a sentence. An **infinitive phrase** is a group of words that includes an infinitive and other words that complete its meaning. Infinitives can be confused with prepositional phrases. Remember that infinitives end with verb forms. Prepositional phrases end with nouns or pronouns.

EXAMPLES

Infinitive phrase: *To write novels* is hard work.
Infinitive as a subject: *To write* is Nathan's lifelong dream.
Infinitive as a predicate noun: Nathan's plan is *to study*.
Infinitive as a direct object: He refuses *to quit*.

Identifying Infinitives

Directions Draw a line under each infinitive used as a noun in the sentences below.

1. The main thing Nathan wants is to write novels.

2. He needs to read many different types of books.

3. An important skill for a writer is to observe.

4. To notice details in everyday life is the first step.

5. Nathan has always liked to read.

6. As a writer, he hopes to improve.

7. He wants to attend a writing workshop.

8. Mrs. Riley offered to help Nathan with his writing.

Take Note

Directions **Choose words from the box below that match each definition. Write them on the lines.**

convince	edit	inaccurate	irresponsible	journalists
persuasive	support	translation	indifferent	viewpoint

1. Reporters are these ____________________

2. Careless or negligent ____________________

3. Mistaken or wrong ____________________

4. To correct or check over written materials before printing ____________________

5. Words put into another language ____________________

Make Your Case

Directions **Read the article. Use words from the box above to fill in the blanks.**

When you write a **(6)** ____________________ paper, you are trying to get someone to support your **(7)** ____________________. In this type of writing, it is important to keep your ideas clear so they will **(8)** ____________________ others of something. If you give good reasons and examples that **(9)** ____________________ your idea, you may sway your readers to adopt your position. If your reasons are weak, and you do not give strong examples, your readers may be unaffected by your writing and remain **(10)** ____________________.

Writing Fractions as Decimals

REMEMBER

All fractions can be written as decimals. To write a fraction as a decimal, divide the numerator by the denominator. A **non-terminating** decimal repeats groups of 1 or more digits forever. To show a non-terminating decimal, draw a line over the digits that repeat. After a certain point in a **terminating decimal,** all digits are zeros.

EXAMPLES

Terminating decimal

$\frac{1}{2} = 2\overline{)1} = 2\overline{)1.0}$ with quotient 0.5

Non-terminating decimal

$\frac{1}{6} = 6\overline{)1} = 6\overline{)1.0}$ with quotient $0.1\overline{666}$

Make It a Decimal

Directions **Write each fraction as a terminating or non-terminating decimal. Show your work.**

1. $\frac{1}{3}$ **2.** $\frac{3}{4}$ **3.** $\frac{5}{8}$ **4.** $\frac{5}{6}$

5. $\frac{2}{3}$ **6.** $\frac{4}{5}$ **7.** $\frac{2}{9}$ **8.** $\frac{7}{12}$

Book Ends

Directions **Look at the surface of Jamal's book below. Then use the information to complete the page.**

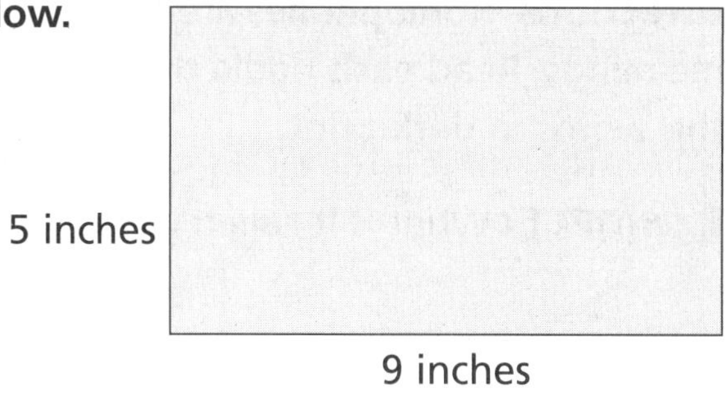

1. Each dimension of the book's surface is rounded to the nearest inch.

   a. What is the least possible length and width the book can have?

   ____

   ____

   b. What is the greatest possible length and width the book can have?

   ____

   ____

2. Use your answers from Problem 1a.

   a. Find the least possible perimeter the book can have. ________

   b. Find the greatest possible perimeter the book can have. ________

3. Use your answers from Problem 1b.

   a. Find the least possible area the book can have. ________

   b. Find the greatest possible area the book can have. ________

Word Pairs

Directions Homophones are words that sound alike but have different spellings and meanings. Read each riddle then solve it using homophone word pairs. Hint: Think about the words in dark print.

EXAMPLE: What is it called when you write correctly?

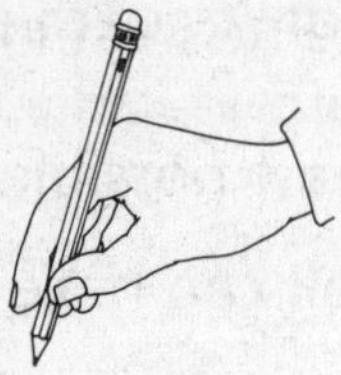

right ______ write ______

1. What do you call **stationery** that is never moved?

______ ______

2. What do you call a book about different kinds of **reed** instruments?

______ ______

3. What do you call a **council** of people that gives advice?

______ ______

4. What is it called when you discover you've been **fined** for an overdue library book?

______ ______

5. What do you call a fictional story about a donkey that loses its **tail**?

______ ______

6. What do you call a poem about money **owed** to a friend?

______ ______

America's Poet

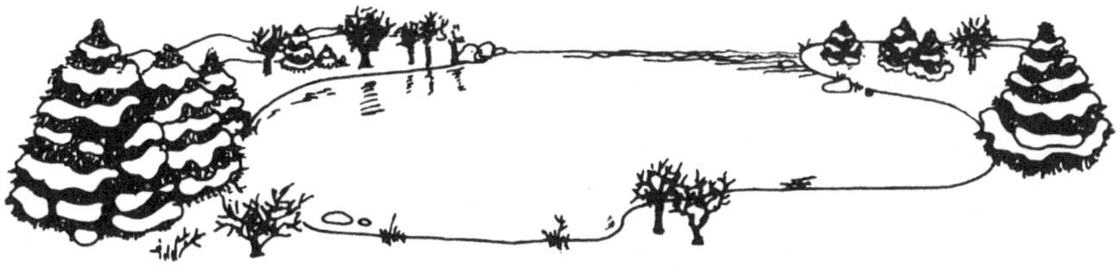

Robert Frost, one of America's best-loved poets, was born in San Francisco in 1874. When Frost was eleven, his father died, and the family moved to Massachusetts. After attending college, Frost settled on a farm in New Hampshire with his wife, Elinor. Some of Frost's best poems were written while running the farm and raising a family.

In 1912, Frost moved to England with his family. In England, he finally earned recognition as a poet when he published his first collection of poems, *A Boy's Will*. His second collection, *North of Boston*, followed shortly after. When Frost and his family returned to America in 1915, Frost was an established poet.

Frost continued to publish his poetry. Most of his poems were about country life and nature—farms, woods, mountains, and the people who lived in the country.

Directions Circle the letter of the best answer.

1. Robert Frost wrote poems about all of the following except

A. farms. **B.** woods.

C. mountains. **D.** factories.

2. What did Robert Frost do right after he attended college?

A. He moved to England. **B.** He moved to a farm.

C. He moved to Massachusetts. **D.** He moved to San Francisco.

3. How did Robert Frost's life change in England?

A. He became well-known as a poet.

B. He became very wealthy.

C. He taught at the university level.

D. He got married.

4. Which of the following words from the passage is a gerund?

A. settled **B.** attending

C. earned **D.** established

CHAPTER 5

Party Politics

Throughout United States history, political parties have dominated elections. A political party is a group of people who share similar ideas about government. Parties try to gain control of the government. They raise money to support their candidates and try to motivate people to vote in elections.

Although there is no constitutional requirement for political parties, their rise is inevitable in a democracy. After George Washington took office as our nation's first president, two views of the country's future emerged. Treasury Secretary Alexander Hamilton favored a strong federal government. The Federalists, as they were known, pushed for a strong military and close ties with England. Many of the country's wealthiest businessmen and professionals supported Hamilton's ideas.

Followers of Secretary of State Thomas Jefferson called themselves Democratic-Republicans. They opposed a strong central government, which they felt intruded on states' rights. This party attracted many rural voters who also supported free trade and low government spending. Today's Democratic party traces its roots to Thomas Jefferson and his followers. During the 1820s, this party expanded nationally, increasing its support by attracting people from the cities as well.

These parties evolved to meet new challenges. Northern Republicans controlled the national government after the Civil War. The Democrats gained strong support in the South. Until recently, the South remained strongly Democratic. Under Democratic Presidents Franklin Roosevelt, John Kennedy, and Lyndon Johnson, the federal government's role in social programs expanded.

The Whig party developed in the 1830s to oppose Democratic President Andrew Jackson. The Whig party favored national unity, protective tariffs, and limits on presidential power. In 1840, they elected William Henry Harrison as president. They lost most of their support by 1860, however, because they were unwilling to take a strong stand against slavery.

Antislavery groups created the Republican Party in 1854. Republicans also supported protective tariffs and full citizenship for immigrants. After the Civil War, Republicans supported rebuilding the South and extending voting and civil rights. The Republican Party, or the Grand Old Party (GOP), as it is often called, has also changed over the years. In the past twenty years, it has supported several business policies to promote economic growth.

Today, Democrats and Republicans continue to debate on issues such as the federal budget, foreign policy, social programs, and the environment. They have also responded to a growing interest in third parties. Regardless of their differences, both parties stand united in protecting America's democracy.

Party Politics

Directions Using what you have just read, answer the questions.

1. What is this article mainly about?

2. What are the two main political parties in America?

3. Why do you think the Federalists wanted strong ties with England?

4. Why do you think more rural voters were attracted to a party that supported states' rights?

5. How are the Democratic and Republican parties alike? How are they different?

6. Find a sentence in this article that is an opinion and write it on the lines.

Phrases

REMEMBER

A **prepositional phrase** is a group of words that begin with a preposition and end with a noun or a pronoun acting as the object of the preposition.
An **adjective phrase** is a prepositional phrase used to modify a noun or a pronoun. An adjective phrase usually modifies the noun or pronoun directly in front of it.
An **adverb phrase** is a prepositional phrase that is used to modify a verb, adjective, or adverb.

EXAMPLES

Prepositional phrase: The American flag is recognized *around the world*.
Adjective phrase: Stars represented each new state *in the Union*.
Adverb phrase: The flag's design has changed *over the years*.

What's That Phrase?

Directions Draw a line under each prepositional phrase in the sentences. Then write if the phrase is an adjective phrase or an adverb phrase on the lines.

1. The flags continued to have a variety of designs. ____________
2. Some versions had stars arranged in geometric patterns. ____________
3. The flag was created to identify patriot ships during the American Revolution. ____________
4. President Eisenhower fixed the arrangement of the flag's stars and stripes. ____________
5. The flag is the living symbol of our country. ____________
6. The "Betsy Ross" flag is surrounded by many myths. ____________
7. Some historians think that Francis Hopkinson is the designer of the first American flag. ____________
8. The original "star-spangled banner" hangs in a Washington, D.C. museum. ____________

Know Your Rights

Directions **Choose words from the box below that match each definition. Write them on the lines.**

Constitution	contributor	democracy	envisioned	guaranteed
government	vote	monarchy	unlawful	veto

1. A donor or supplier ____________________

2. Illegal or criminal ____________________

3. Reject or refuse ____________________

4. A form of government in which people vote ____________________

The Founding Fathers

Directions **Read the article. Use words from the box above to fill in the blanks.**

When the founding fathers wrote the **(5)** ____________________, they **(6)** ____________________ a new kind of government. They wanted to be sure that government leaders would never be able to abuse their power. This did away with having a **(7)** ____________________, or one supreme leader. This new **(8)** ____________________ gave the American people the power to choose their leaders. Many other countries do not have the ability to **(9)** ____________________ for a president. These countries are not a **(10)** ____________________, or government by the people, like the United States.

Using a Grid

REMEMBER

A **grid** is a plane formed by an *x*-axis and a *y*-axis. A **coordinate pair** identifies the location of a point on a grid. The first number in a coordinate pair tells the location on the *x*-axis. The second number tells the location on the *y*-axis.

EXAMPLE

Point A is located at the coordinate (3, $^-2$).

x-axis   *y*-axis

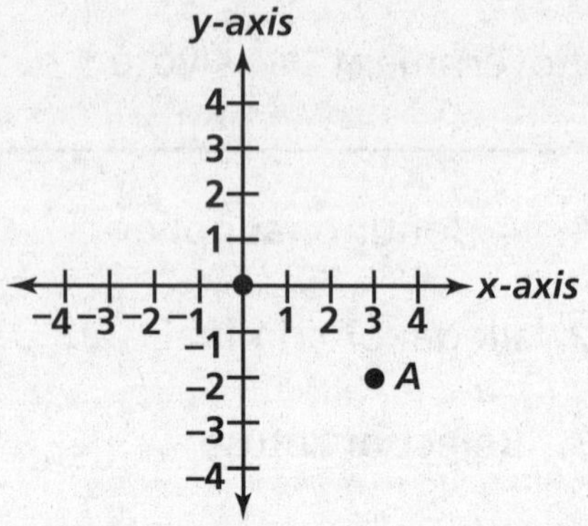

To the Point

Directions Write the coordinates of each point.

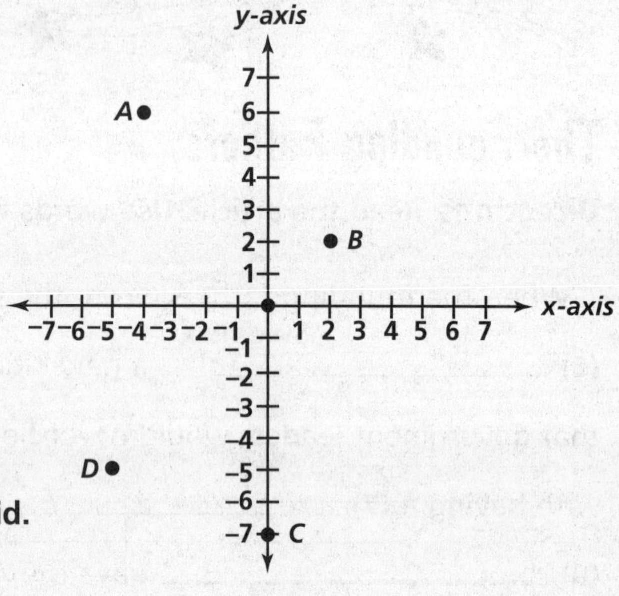

1. *A* ___________
2. *B* ___________
3. *C* ___________
4. *D* ___________

Directions Draw and label each point on the grid.

5. *E* ($^-6$, 4)
6. *F* (5, 1)
7. *G* (5, $^-4$)
8. *H* ($^-1$, $^-3$)

Forgetful Fundraiser

Directions **Use the information below to complete the page.**

A local political group was selling T-shirts to raise money for a candidate. T-shirts sold for $8.50 and sweatshirts for $14.75. After Saturday's sale, volunteers had sold a total of 10 shirts for $110.00. Unfortunately, no one kept a record of each shirt sold.

1. What would the total be if all the shirts sold were

a. T-shirts? __________

b. Sweatshirts? __________

2. Guess how many

a. T-shirts were sold? __________

b. Sweatshirts were sold? __________

3. Explain how you would check your guess.

____

4. Complete the table to show how much money would be collected for each possible combination of 10 shirts sold.

T-Shirts	Sweatshirts	Total Amount
0	10	$(0 \times \$8.50) + (10 \times \$14.75) = \$147.50$

5. How many of each type of shirt was actually sold?

a. T-shirts = __________

b. Sweatshirts = __________

Let Freedom Ring!

Directions Find the words listed in the box. Then circle them in the puzzle. The words can be hidden across, down, or diagonally.

democracy	democrat	election	executive	judicial	laws
legislative	republican	rights	tax	veto	vote

h	e	d	e	m	e	c	v	a	c	e	r
i	x	a	u	l	c	l	e	r	o	l	e
l	e	g	i	s	l	a	t	i	v	e	p
m	c	p	x	z	l	b	o	g	s	c	u
j	u	d	i	c	i	a	l	h	w	t	b
g	t	y	f	t	p	s	w	t	v	i	l
r	i	r	n	e	r	y	g	s	t	o	i
a	v	v	q	a	t	i	n	a	u	n	c
d	e	m	o	c	r	a	c	y	t	d	a
t	n	e	b	t	u	m	x	t	i	b	n
r	r	o	r	d	e	m	o	c	r	a	t
y	n	z	e	a	w	n	l	e	c	p	e

Coordinate Grids

Directions Plot and label the following points on the coordinate grid.

1. A (7, 2)
2. B (−5, 2)
3. C (7, −4)
4. D (−5, −4)

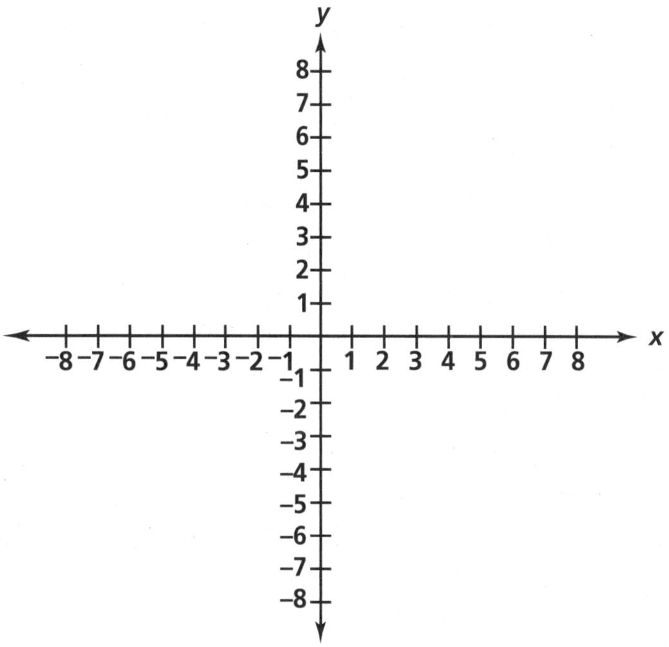

5. What shape is formed by connecting the points?
6. What is the length of this shape?
7. What is the width of this shape?
8. What is the area, in square units, of this shape?

Directions On the coordinate grid below, plot a shape that has an area of exactly 10 square units.

9.

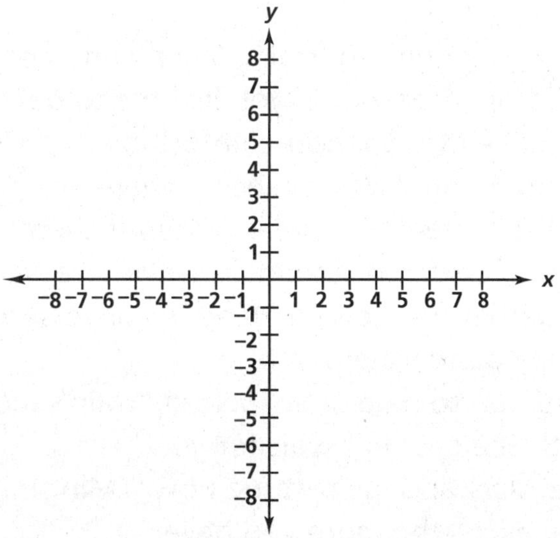

CHAPTER 6

Min's Good Fortune

z"Let me read mine now, Mom!" cried Quan, pushing away his empty plate. "A stitch in time saves nine."

"That's easy," laughed Mrs. Li. "Do your homework when it's assigned to save yourself a lot of worry in the future. Min, what does yours say?"

Min looked around the table at everyone. It was a family tradition to celebrate birthdays at the local Chinese restaurant, and today Min was thirteen years old. It was also tradition to share your fortune with everyone. "I want to keep mine a secret until I've figured out what it means," Min said. This one meant something special.

Min woke up on Saturday morning. The sun was shining through the windows. Today Min, Carol, and some other friends had planned a picnic. She was very excited.

"Min, are you awake?" Her mother's voice roused her from her daydreams. "Get up, hon, I want you to run an errand."

"Oh, Mom, what is it?" Min dragged herself into the kitchen.

"Do me a favor and take these groceries to Mrs. Spencer. She hurt her back and can't go to the store right now." Min got ready, picked up the groceries, and dashed across the street. She knocked impatiently on Mrs. Spencer's back door, then went in. Min placed the brown paper bag on the kitchen table and peaked into the living room. The drapes were still closed and the room was dark. Min walked across to the hallway and called, "Mrs. Spencer, it's Min. Mom sent over some groceries."

A faint voice answered from upstairs. "Thank you, Min. I'm sorry I can't come down to greet you."

"Mrs. Spencer, are you all right? May I come up?" Min asked.

From the bedroom doorway, Min smiled shyly at Mrs. Spencer. She looked so fragile. "Mrs. Spencer, you don't look well. Can I fix you some breakfast?" The look on Mrs. Spencer's face showed this was a welcome offer. "Oh well," Min thought, "Carol can wait half an hour."

Min made breakfast and carried it upstairs. She planned on leaving it and heading home. Then Min asked, "Mrs. Spencer, would you like me to get your mail?" Next, she plumped up the pillows to make Mrs. Spencer comfortable. As Min put the mail next to Mrs. Spencer, she noticed something. "Mrs. Spencer, where are your glasses?"

"I broke them when I fell. I feel so helpless without them!"

Hours later, Mrs. Li came across the street to see where Min was. She found Min and Mrs. Spencer sitting together on the bed, deep in conversation. Opened letters and discarded newspapers lay beside them.

"It was really sweet of you to help Mrs. Spencer," Min's mother said as they headed home. "Carol called. She couldn't wait for you."

"It's alright, Mom. I understand my fortune now." Min smiled and recited it to herself: "The more you give away, the more you have."

Min's Good Fortune

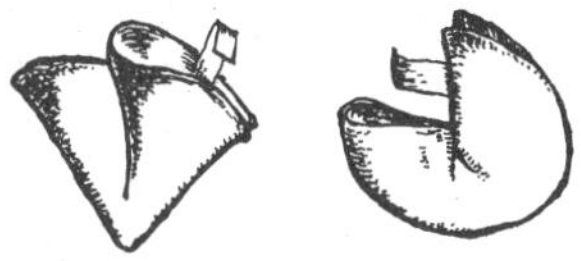

Directions **Using what you have just read, answer the questions.**

1. Why did Mrs. Li send groceries to Mrs. Spencer?

2. What happened after Min fixed breakfast for Mrs. Spencer?

3. Why did Min stay and help Mrs. Spencer?

4. What did Min's fortune mean?

5. Do you think Min was upset she missed the picnic? Why or why not?

6. Suppose Min's fortune was: *Life is a picnic.* Do you think she would have acted differently? Why or why not?

Independent and Dependent Clauses

REMEMBER

An **independent clause** can stand alone as a sentence because it expresses a complete thought. Independent clauses can be joined by commas and coordinating conjunctions such as *for*, *and*, *not*, *but*, *or*, *yet*, and *so*. They can also be joined by independent clauses.

A **dependent clause** cannot stand alone as a sentence because it does not express a complete thought. Dependent clauses can be introduced by words such as *although*, *as*, *because*, *before*, *if*, *since*, *so*, *though*, *unless*, *until*, *when*, *wherever*, and *while*. Dependent clauses can come at the beginning or end of a sentence. A comma is used after a dependent clause that comes at the beginning of a sentence.

EXAMPLES

Independent clause: *Kevin and Roland wanted to make a pizza*, but *they didn't have a recipe*.

Dependent clause: *Although they had ingredients for the toppings*, they needed a pizza crust.

Identify the Clause

Directions Label each underlined clause as **independent** or **dependent**.

1. The boys could make the pizza dough themselves, or they could buy a prepared crust. __________
2. Roland and Alex bought a ready-made crust because it was easier. __________
3. They put all the ingredients on the counter so they could plan the steps. __________
4. Roland sliced the vegetables, and Kevin made the sauce. __________
5. After Kevin spooned sauce on the dough, Roland added toppings. __________
6. Kevin put the pizza in the oven, and then he set the timer. __________
7. While the pizza cooked, the boys washed the dishes. __________

Edible Explanations

Directions **Choose words from the box below that match each definition. Write them on the lines.**

aroma	awakened	chef	consume	custom
extinguished	misbehaved	originated	behaved	whiff

1. To eat ______________________

2. A professional cook ______________________

3. A pleasant smell ______________________

4. A sniff or breath ______________________

Birthday Customs

Directions **Read the article. Use words from the box above to fill in the blanks.**

Did you know that most of our birthday customs **(5)** ______________________ in the Middle Ages in Germany? The birthday child was **(6)** ______________________ at dawn when the family brought a cake with lighted candles into the bedroom. The family kept the candles lit all day long. After the child's favorite meal was served, it was time for gifts and wishes. The candles had to all be **(7)** ______________________ in one breath or the child's secret wish would not come true. One German **(8)** ______________________ most of us did not adopt was the arrival of the birthday elf who brought extra gifts to well-**(9)** ______________________ children, but brought nothing to children who had **(10)** ______________________ that year.

Symmetry

REMEMBER

Symmetry has to do with how things look. When you fold a figure along a line and the two parts of the figure match exactly, the line is called a **line of symmetry**. Some figures have only one line of symmetry. Some figures have many lines of symmetry.

EXAMPLES

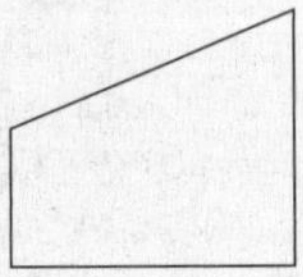

0 lines of symmetry

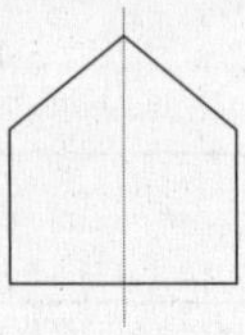

1 line of symmetry

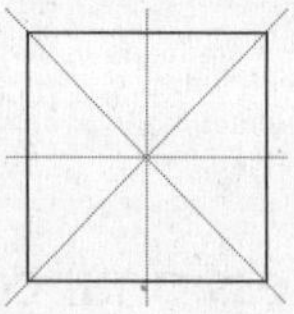

4 lines of symmetry

Drawing the Line

Directions Draw all the lines of symmetry on each figure. Then write the total number of lines of symmetry.

1. 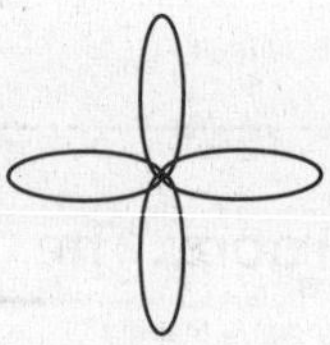

______

2. 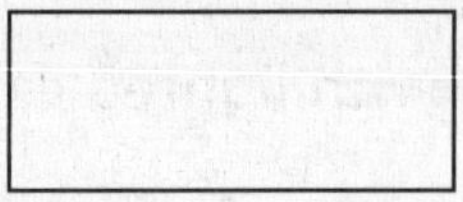

______

3. 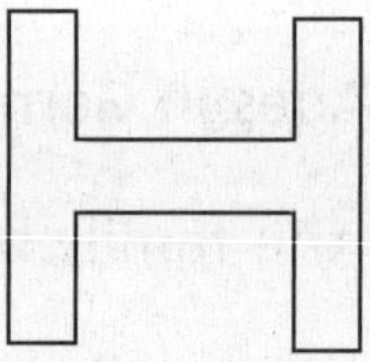

______

4. 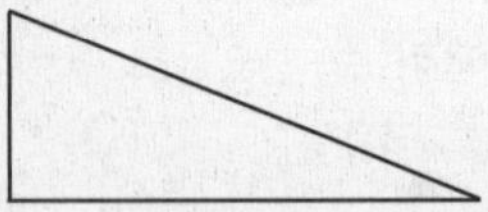

______

5. 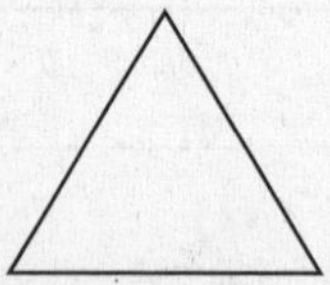

______

6. 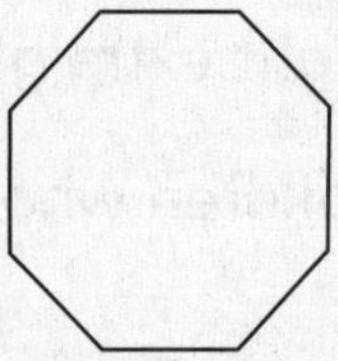

______

Circle Graphs

Directions Students at Jefferson School made a circle graph showing their favorite desserts. Every student in the school was included in the poll. Use the graph below to complete the page.

Favorite Desserts

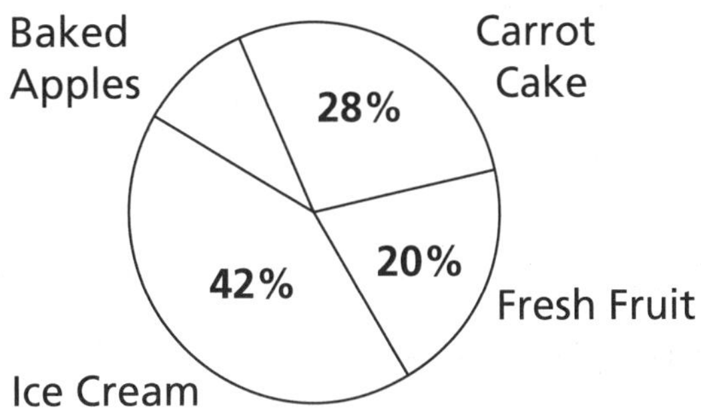

1. What percent of students chose carrot cake as their favorite dessert? __________

2. Write this percent as a fraction. Reduce to lowest terms. __________

3. Write this percent as a decimal. __________

4. If 190 students chose fresh fruit as their favorite dessert, how many students attend Jefferson School? __________

5. What percent of students chose baked apples as their favorite dessert? (This percent is not written on the graph.) __________

6. Forty percent of the students who chose baked apples like it with a caramel glaze. How many students prefer baked apples with caramel glaze as their favorite dessert? __________

7. How many students prefer baked apples without caramel glaze as their favorite dessert? __________

Kitchen Quiz

Directions **Read each riddle and solve it using homophone word pairs.**
Hint: Think about the words in dark print.

EXAMPLE: What do you call a vegetable made out of one **carat** of gold?

A ___carat___ ___carrot___

1. What is the best tool to **grate** food?

__________ __________

2. What kind of animal makes bread **dough**?

__________ __________

3. What do people do when they leave before eating **dessert**?

__________ __________

4. What do you call the doughnut **hole** that someone ate in one bite?

__________ __________

5. What do you call the time during which four couples **ate** dinner?

__________ __________

6. What do you call the step in a recipe in which **thyme** seasoning is added?

__________ __________

7. What name did the chef give to a dessert in which a **berry** was covered in chocolate?

__________ __________

Foods That Provide Calcium

Calcium is essential for the formation and maintenance of strong bones and teeth. Your body needs calcium to ensure that muscles and nerves function properly.

Milk and other dairy products are excellent sources of calcium. Vegetables such as broccoli, kale, collard greens, turnips, and Chinese cabbage also contain calcium. In addition, some kinds of fish, nuts, and dried beans are good dietary sources of this important mineral.

Just as there are foods that contain calcium, there are also foods that can cause your body to lose calcium. For example, caffeine—which is found in coffee, tea, and many soft drinks—can prevent your body from absorbing calcium. Try to avoid drinking these beverages, so your body can utilize all the calcium it gets.

Directions **Circle the letter of the best answer.**

1. What is the main idea of this passage?

A. Your body needs calcium, and there are many foods that have it.

B. Milk and dairy products are a great source of calcium.

C. Many vegetables contain calcium.

D. Some foods prevent your body from absorbing calcium.

2. Which of the following sentences is a complex sentence?

A. Milk and other dairy products are excellent sources of calcium.

B. Calcium is essential for the formation and maintenance of strong bones and teeth.

C. Try to avoid drinking these beverages, so your body can utilize all the calcium it gets.

D. Your body needs calcium to ensure that muscles and nerves function properly.

3. Which two words in the passage have almost the same meaning?

A. formation and maintenance

B. essential and important

C. prevent and avoid

D. utilize and cause

CHAPTER 7

Susan B. Anthony

Susan B. Anthony
1820–1906

Susan B. Anthony was born in 1820 and grew up in Adams, Massachusetts. She lived during a time when women did not enjoy many of the rights they have today. Susan's mother ran the household and Susan worked with her father in his mill. He was the first mill owner in the area to employ females. However, women were not allowed to hold supervisory positions and they didn't earn the same amount of money as men.

When Susan was 18, her father lost his mill. Susan and her sisters needed to work to help support the family. Susan chose to teach. During these years, many men asked her to marry them. Susan always turned down their proposals. She did not agree with the laws for married women. If a woman in Susan's time was married, she had few rights. She couldn't own property or even keep the money she earned if her husband didn't approve. Some men even went unpunished for abusing their wives. Susan told her friends and family, "I will not be any man's slave."

While teaching, Susan became very angered with the lack of women's rights. Slavery was still legal, and this also appalled Susan. She met with other people to discuss these issues.

One of these people was Frederick Douglas, an ex-slave who worked to end slavery in the United States. She also met with Elizabeth Cady Stanton. Mrs. Stanton and Susan formed the first women's rights organization in the United States. It became known as the Suffrage Movement. They fought for equal rights for both African Americans and women. They made speeches along the East Coast, often receiving jeers from both men and women. They spoke about the importance of giving women the right to vote. Many people disagreed with them and even called them evil.

This abuse didn't slow Susan down. In 1865, part of this fight was won with the passage of the Thirteenth Amendment to the Constitution, which made slavery illegal in the United States. Three years later, the Fourteenth Amendment granted all men, regardless of race or ethnicity, the right to vote. Although she was pleased that African Americans won freedom and the right to vote, Susan wouldn't rest until women could also vote.

In 1872, some lawyers claimed that women were also granted the right to vote with the Fourteenth Amendment. If the words of the amendment were interpreted differently, a case could be made that women, being citizens, had the right to vote. On November 5, 1872, Susan and fifteen other women voted in an election. She was arrested twenty-three days later for voting illegally. Susan was fined $100, however, she never paid the fine. For the rest of her life, Susan worked for equal rights for women. In 1920, nearly twenty years after her death, Congress passed the Nineteenth Amendment, which gave women the right to vote.

Susan B. Anthony

Directions Using what you have just read, answer the questions.

1. What is this article mostly about?

2. Why did Susan say she would never marry?

3. What was the Suffrage Movement?

4. Why do you think some men and women did not accept Susan's ideas?

5. Which amendment gave women the right to vote?

6. How have the lives of women today changed from those in Susan B. Anthony's time?

Nonrestrictive Clauses

REMEMBER

A **nonrestrictive clause** is an adjective clause that is not essential to the word that it modifies. It supplies additional information that can be removed from the sentence without changing the basic meaning. Commas are used to set off a nonrestrictive clause.

EXAMPLE

The actor Mel Gibson, *who was born in New York*, grew up in Australia.

Clause Pause

Directions **Draw a line under each nonrestrictive clause in the sentences below.**

1. After graduating from high school, Gibson took his first job, which was in a juice factory.
2. Mel's sister, who recognized his acting potential, sent an application for him to an acting school.
3. Gibson's acceptance to the school, which came as a surprise, changed his life.
4. *Mad Max*, which was Gibson's first movie role, was a huge success.
5. Gibson fell in love with Robyn Moore, whom he married in 1980.
6. He played the part of Shakespeare's Hamlet, who seeks revenge for his father's murder.
7. Later he starred in *Braveheart*, which is a film about a Scottish rebel.
8. Gibson also directed the film, which earned awards for Best Director and Best Picture of 1995.

Writer's Words

Directions Choose words from the box below that match each definition. Write them on the lines.

author	biography	celebrity	challenging	contemporary
diligent	libel	nonfiction	perspective	research

1. A point of view or outlook ______________________

2. Hardworking, thorough ______________________

3. A famous person, superstar ______________________

Story of My Life

Directions Read the article. Use words from the box above to fill in the blanks.

A **(4)** ______________________ is a story in which the **(5)** ______________________ writes about another person's life. All of these types of stories are **(6)** ______________________, or true. Some stories tell about people who lived long ago. A biographical author must carefully **(7)** ______________________ letters and works of art, like paintings completed during that time period to find out all the details of the person's life. Some stories tell about modern, or **(8)** ______________________ people. Researching information about people living today can be just as **(9)** ______________________. The author should have permission from the person he or she is writing about. If not, the author can be sued for **(10)** ______________________, or sharing false information.

Similar Figures

REMEMBER

Similar figures have the same shape but may differ in size.

EXAMPLE

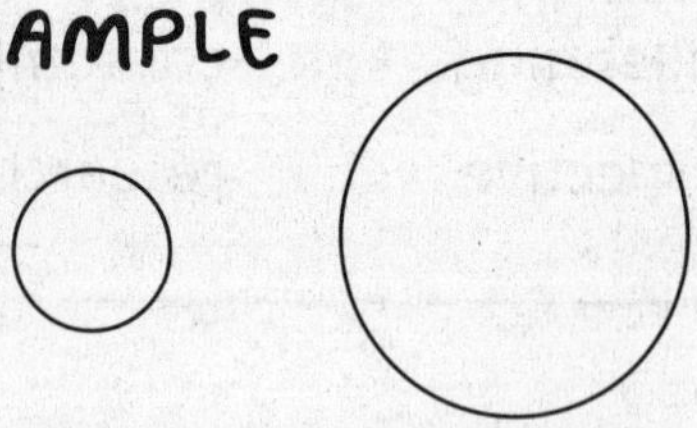

All circles are similar.

Figure Match Up

Directions Circle the two figures that are similar to the first one shown.

1.  a.  b. 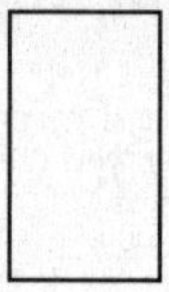 c. 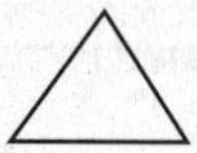 d. 

2.  a.  b.  c.  d. 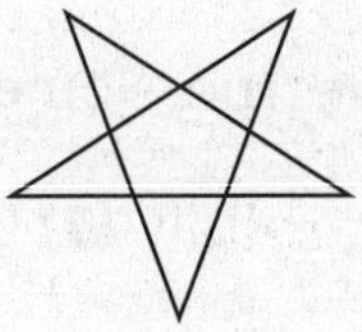

3. 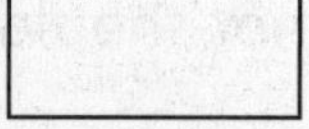 a.  b.  c.  d. 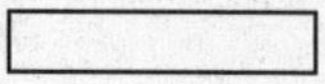

4. 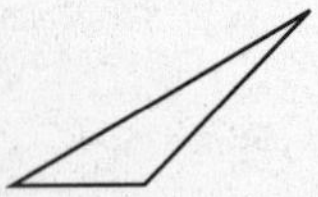 a. 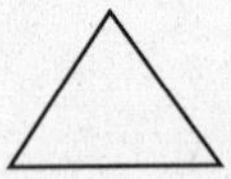 b. 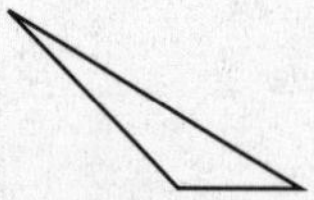 c. 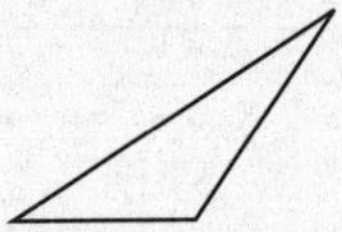 d. 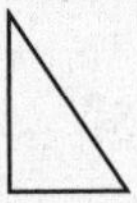

Heads or Tails?

Directions **A penny has two sides—heads and tails. Use this information to complete the page.**

1. Suppose you toss 3 pennies into the air and they land on the floor. Complete the table to show the combinations of flips that are possible. The first one is done for you.

Penny 1	**Penny 2**	**Penny 3**
heads	heads	heads

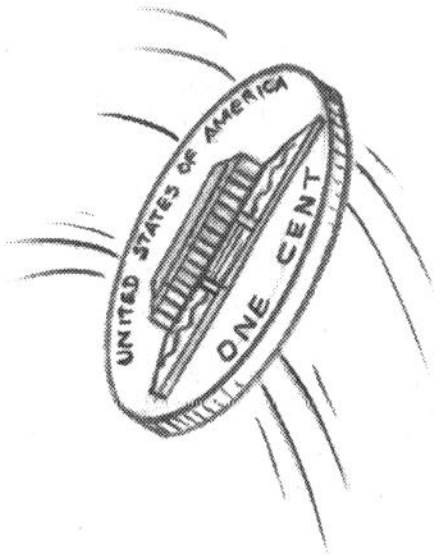

2. How many combinations of flips are possible? __________

3. What is the probability that all three pennies will land with the same side up—either 3 heads up or 3 tails up? Explain.

____

____

4. How many times as likely are the 3 pennies to land with 2 heads up and 1 tails up than to land with 3 heads up? Circle your answer.

two times three times four times

5. Explain your answer.

____

____

Celebrity Scramble

Directions Read each definition. Unscramble the letters to find the careers. Write the words on the lines. Then use the numbered letters to solve the riddle.

1. cendar A person who moves to music

___ ___ ___ ___ ___ ___
(1 under letter 2, 2 under letter 3)

2. rontastua A person who travels in space

___ ___ ___ ___ ___ ___ ___ ___ ___
(3 under letter 8, 4 under letter 9)

3. rocta A person who plays a part in a movie or play

___ ___ ___ ___ ___
(5 under letter 4)

4. abbre A person who cuts hair

___ ___ ___ ___ ___ ___
(6 under letter 4)

5. tenssitic A person who studies things related to the world

___ ___ ___ ___ ___ ___ ___ ___ ___
(7 under letter 3)

6. genirs A person who entertains with songs

___ ___ ___ ___ ___ ___
(8 under letter 4, 9 under letter 6)

7. preorrte A person on the TV, radio, or in a newspaper who informs people of the latest news

___ ___ ___ ___ ___ ___ ___ ___
(10 under letter 3)

8. thruoa A person who writes a book

___ ___ ___ ___ ___ ___
(11 under letter 4)

9. ralwey A person who tries a case in court

___ ___ ___ ___ ___ ___
(12 under letter 4)

NOW SOLVE THE RIDDLE:

What kind of book does a racecar driver like to read?

___ ___   ___ ___ ___ ___ ___ ___ ___ ___ ___ ___ ___ ___ ___

1 2   1 3 4 5 6 7 5 8 9 1 10 11 12

Possible Outcomes

Directions A six-sided number cube is rolled three times. Answer the following questions:

1. How many combinations of rolls are possible?

____

2. What is the probability of rolling the same number all three rolls?

____

____

3. Is it more likely to roll the same number three times or a different number each time? Explain your answer.

____

____

____

CHAPTER 8

Shoes Are My Life

When I head off to work in the morning, I don't just think of myself as a shoe salesman. I think of myself as the guy who gets the world moving. If it weren't for my eye for detail and knowing what the customer wants, there might be several people running around out there shoeless!

Okay, so maybe that's not entirely true, but I'm certainly not just peddling shoes. There's a lot more that goes into selling shoes. Feet are sometimes a very "touchy" subject. You have to approach outfitting the foot very carefully. First, you have to learn what the customers want from their shoes and how they want their shoes to make them feel. Then you need to find out where the customer is going to wear the shoes. Are they for the office, the home, or going out dancing? The shoes, of course, need to fit every activity that person is planning to do. You don't want to wear a pair of stiff loafers all day if you're going to spend all of your time on your feet. Fashion may lead them in, but comfort makes the sale.

I can always tell the customers who need my help the most. They walk in, sometimes limping and wincing. They're usually very cautious because of a bad shoe experience in the past. Watching them walk out with full, confident steps and a new pair of shoes is what makes this job worthwhile.

Shoes weren't always my game, though. I've tried to sell sports equipment, outdoor gear, and even flowers. No other job gives me the enjoyment I get in knowing that I'm helping people get moving. I guess you can say that I put both my heart and "sole" into my job!

Shoes Are My Life

Directions **Using what you have just read, answer the questions.**

1. The narrator lists three things to keep in mind when helping a customer pick out shoes. List two of them.

____

____

2. Why do you think it is important that a person's shoes fit the activities they are planning to do?

____

____

3. What does the sentence "Fashion may lead them in, but comfort makes the sale" mean?

____

____

4. What would be another good title for this selection?

____

5. The narrator makes the statement that he helps people get moving. What does he mean by this?

____

____

6. Does the narrator like his job as a shoe salesman? Explain your answer.

____

____

____

7. Why did the author write this selection?

____

____

Dependent Clauses

REMEMBER

A **noun clause** is a dependent clause that is used as a noun. Noun clauses usually begin with words such as *how, that, what, whatever, when, where, which, who, whom, whomever, whose,* or *why*.

An **adjective clause** is a dependent clause that is used as an adjective to modify, or describe, a noun or a pronoun. Adjective clauses usually begin with the relative pronouns *that, which, who, whom,* or *whose*.

An **adverb clause** is a dependent clause that is used as an adverb to modify a verb, adjective, or adverb. Adverb clauses usually begin with subordinating conjunctions such as *although, because, if, since, though, until,* or *when*.

EXAMPLES

Noun clause: Lucy's reason for not cleaning her room is *that it's too messy*.

Adjective clause: Cleaning a messy room is a job *that can be overwhelming*.

Adverb clause: You can do the job more efficiently *if you have a plan*.

Classifying Clauses

Directions Draw a line under each dependent clause in the sentences below. Then write if the clause is a noun clause, an adjective clause, or an adverb clause on the lines.

1. There are several hints that can make cleaning easier. ____________
2. Ask whomever is around to help you with the task. ____________
3. Use cleaning tools that save time. ____________
4. Before you go to bed, pick up any clutter in your room. ____________
5. Give or throw away what you don't use. ____________
6. Vacuuming is the step that comes last. ____________
7. An unmade bed is what makes a room look messy. ____________
8. Work from top to bottom whenever you're cleaning a room. ____________

Money in the Bank

Directions **Choose words from the box below that match each definition. Write them on the lines.**

adversity	bankrupt	accounts	currency	earnings
tax	market	oversee	wealthiest	

1. Hardship or difficulty ____________________

2. The money system of a country ____________________

3. Having the most money or riches ____________________

4. Broke, penniless, having no money ____________________

Taxing Times

Directions **Read the article. Use words from the box above to fill in the blanks.**

When people work, they have to pay an income **(5)** ____________________ on the money they make. They must report their salary and any other **(6)** ____________________ to the government. People who invest in the stock **(7)** ____________________ must know how much profit was made or lost on each investment they have. Then they must pay taxes on their gains. Banks usually monitor the interest on these savings **(8)** ____________________ each day. The interest adds up and the account holder reports that money as extra income. Often the **(9)** ____________________ taxpayers have many sources of income from of property and investments. They often hire someone to **(10)** ____________________ their finances and to calculate their taxes correctly.

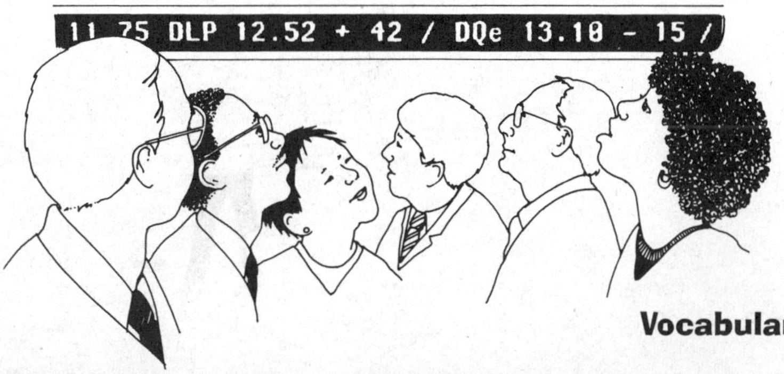

Repeating Patterns

REMEMBER

A **tessellation** is a pattern made by repeating a single shape in a side-by-side manner with no overlapping or gaps. A honeycomb is an example of a tessellation. In a honeycomb, a hexagon is repeated. Many kinds of figures can be used to make tessellations, but some cannot. Circles cannot be used to make a tessellation because they leave gaps.

EXAMPLES

A hexagon makes a tessellation.
Circles do not make a tessellation.

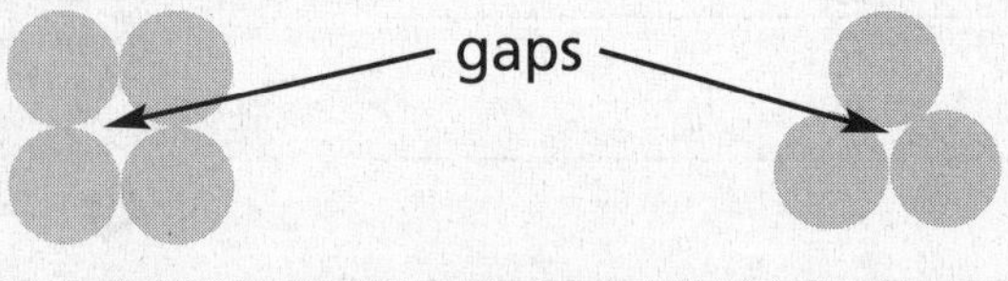

Honeycomb Tessellation

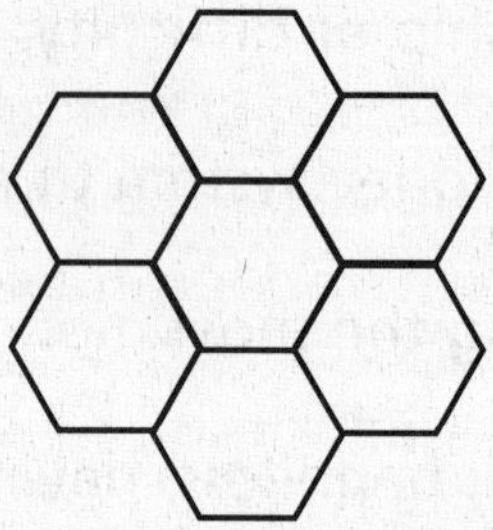

Making Tessellations

Directions Trace the figures on another sheet of paper and cut them out. Turn the cutouts to determine if the figures make tessellations. Circle the figures that can make a tessellation.

1. 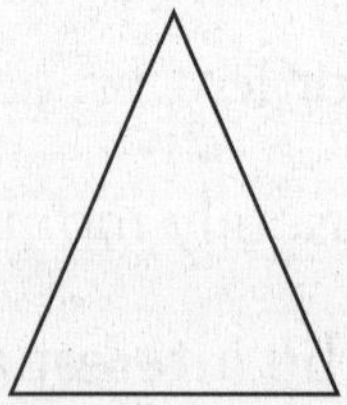

2. 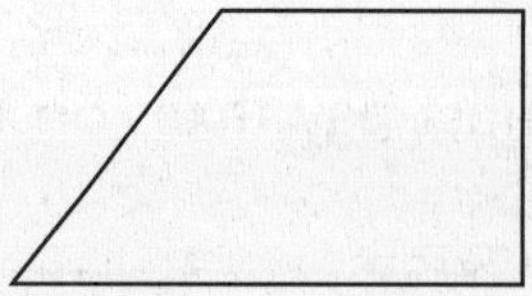

3. 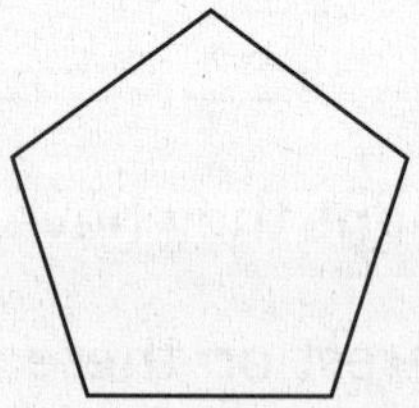

4. 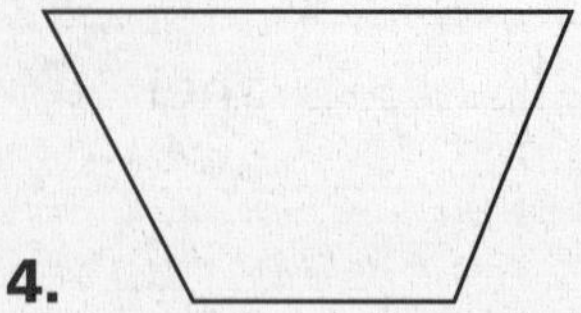

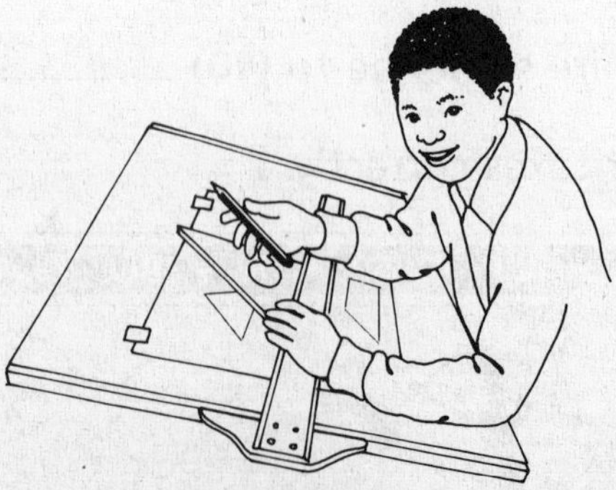

5. 

Painting Time

Directions **Read the information below. Then complete the page.**

Gerry and Keneisha are going to paint a room together. If Gerry works alone, he can paint the room in 6 hours. If Keneisha works alone, she can paint the room in 8 hours.

1. What fraction of the room can Gerry paint in one hour? ____________

2. What fraction of the room can Keneisha paint in one hour? ____________

3. What fraction of the room can Gerry and Keneisha paint in one hour working together? ____________

4. What assumptions are you making in your answer to Problem 3?

____

____

5. Using your answer from Problem 3, determine about how many hours it will take Gerry and Keneisha to paint the room while working together?

____

____

Profession Puzzlers

Directions **Read each riddle and solve it using homophone word pairs. Hint: Think about words that rhyme with those in dark print.**

EXAMPLE: What do you call a boat powered by the wind that's for **sale**?

A ___sail___ ___sale___

1. What do you call a builder who convinces a homeowner to add a **cellar** to her home?

   ______________ ______________

2. What do you call the government charge on a box of **tacks**?

   ______________ ______________

3. What do you call a **bazaar** that has unusual products for sale?

   ______________ ______________

4. What do you call the invoice **billed** to someone for construction work?

   ______________ ______________

5. What do you call a hiding place where you would keep your **cash**?

   ______________ ______________

6. What do you call it when someone discovers they have been **fined**?

   ______________ ______________

Everyone Needs a Dentist!

Tooth and gum diseases are as uncomfortable and unhealthy for animals as they are for humans. For this reason, all around the United States, many dentists work in zoos and parks, cleaning and repairing a variety of fangs, molars, and tusks.

Dentists perform procedures such as cleanings, extractions, and root canals on their animal patients. Since animals usually don't get cavities, dentists rarely do fillings. However, tartar builds up on animals' teeth, just as it does on humans'. Therefore, dentists scrape and clean the animals' teeth. Of course, dentists sedate their animal patients with anesthesia so that they are asleep during the procedures.

Directions Circle the letter of the best answer.

1. Which of the following is **not** true about both animals and humans?

A. Tooth and gum diseases are uncomfortable for animals and humans.

B. Animals and humans get their teeth cleaned.

C. Animals and humans often get cavities.

D. Animals and humans get tartar on their teeth.

2. The word **extraction** means ______.

A. cleaning **B.** removal

C. filling **D.** scraping

3. Why do dentists scrape and clean animals' teeth?

A. to help the animals chew their food

B. to make the animals' teeth whiter

C. to avoid getting bitten

D. to remove tartar

4. How do dentists stay safe when they work on an animal?

A. Dentists wear protective coats. **B.** Dentists carry a weapon.

C. Dentists wear leather gloves. **D.** Dentists sedate the animal.

CHAPTER 9

"Boz"

Charles Dickens was born in 1812, in Portsmouth, England. His family moved to London when he was ten. Unable to go to school because of the high cost, Charles spent his days looking after his younger brothers and sisters.

When Charles was twelve, his father went bankrupt and was sent to debtors' prison. His mother and younger siblings joined his father in prison. Charles was sent to work in a warehouse. There he had the monotonous task of pasting labels on bottles of shoe polish, twelve hours a day, six days a week. After a while, relatives helped out the Dickens family, and Charles was able to attend school.

In his early twenties, Dickens became a newspaper reporter in London. During this time, he sold prose sketches about life in London to local magazines, using the pen name "Boz." He became known through these sketches for the unusual characters, elaborate descriptions, and lighthearted humor.

Dickens began writing monthly magazine installments of his successful first novel, *Posthumous Papers of the Pickwick Club*. For several years, he published novels, such as *Oliver Twist*, in serial-magazine installments before they were finished.

Dickens also became known as a champion of reform. From some of his own experiences, Dickens knew that conditions in public institutions such as schools, hospitals, courts, and prisons were terrible. Through his writing, he sought to improve these conditions. Dickens often used humor to attack social and moral injustices. Later in life he ran several charitable organizations.

Charles Dickens
1812–1870

In 1850, Dickens founded a weekly magazine in which he serialized *A Tale of Two Cities* and *Great Expectations*. Other novels written by Dickens, such as *David Copperfield*, were also published in monthly installments. Some of these novels took nearly two years to complete.

Dickens became world famous and prosperous because of his writing. A natural actor and orator, he often gave public readings of his work to audiences in England and the United States. By the time he died in 1870, Charles Dickens had become one of the most popular English writers.

"Boz"

Directions **Using what you have just read, answer the questions.**

1. What happened to people who went bankrupt in England in the 1800s?

2. What did Dickens do after working in the shoe polish factory?

3. What kinds of sketches did Dickens sell to the newspaper?

4. How did Dickens try to reform conditions in public institutions?

5. How was Dickens' financial situation as a child different than when he was an adult?

6. What is an *orator*?

7. Why do you think that Dickens became a popular writer?

Indirect Objects

REMEMBER

An **indirect object** is a noun or pronoun in the predicate that answers the questions *To whom? For whom? To what?* or *For what?* after an action verb.

EXAMPLES

The drama teacher showed *us* a movie of the play.
(The teacher showed a movie of the play *to whom*? *Us* is the indirect object.)
We gave our *play* a name.
(We gave a name *to what*? *Play* is the indirect object.)

Speaking Indirectly

Directions Underline the indirect object in each sentence.

1. Mr. Brower gave us the addresses of the playwrights.
2. We wrote them concerning their work.
3. Mr. Brower assigned the class a month-long drama project.
4. Evan showed his group an original script.
5. One of the playwrights wrote Mr. Brower a letter about visiting the class.
6. Seth handed Marsha the microphone for the dramatization.
7. Hank, Alonzo, and Jean told us their play idea.
8. Greta loaned Alyssa a cape for the play.
9. The parents' club bought us a new curtain for the stage.
10. The faculty gave four students special awards for their roles in the play.

On the Stage

Directions Choose words from the box below that match each definition. Write them on the lines.

applause	audience	converse	dramatic	entirety
participate	poets	recite	rhythm	stanzas

1. To become actively involved in something ____________________
2. To talk or communicate with another person ____________________
3. Clapping to show approval ____________________
4. People who watch a performance; spectators ____________________

Poetry Night

Directions Read the article. Use words from the box above to fill in the blanks.

Poetry readings allow people to enjoy the sound and **(5)** ____________________ of poems. At some poetry readings, **(6)** ____________________ are invited to read their own work. Other poetry readings share works from poets of a particular country, or focus on the different works from just one poet. Readers may read a poem in its **(7)** ____________________, or they may **(8)** ____________________ just a few **(9)** ____________________, or lines of a longer poem. Some poetry readings are quite **(10)** ____________________ because the reader becomes emotionally involved in the poem.

Learning Angles

REMEMBER

An **angle** is formed by two rays with a common endpoint. The rays that form the angle are called the **sides**. The endpoint is called the **vertex** of the angle. Letters are used to name angles. The symbol $\angle$ stands for an angle. An angle can be named for the vertex letter alone or by the three letters that form the angle. When using three letters, the vertex must be the middle letter.

EXAMPLE

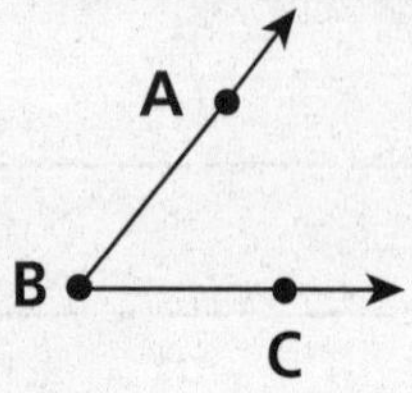

$\angle ABC$, $\angle CAB$, or $\angle B$

What's Your Angle?

Directions Write three names for each angle.

1.

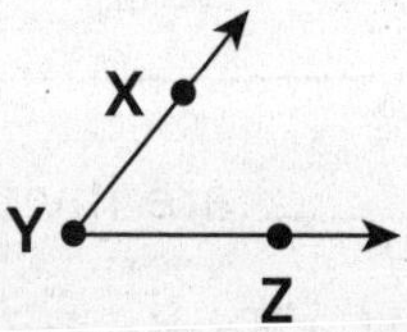

________ ________ ________

2.

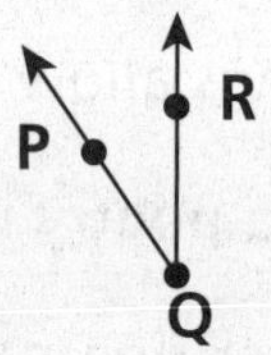

________ ________ ________

3.

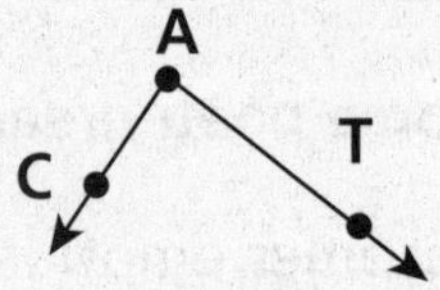

________ ________ ________

4.

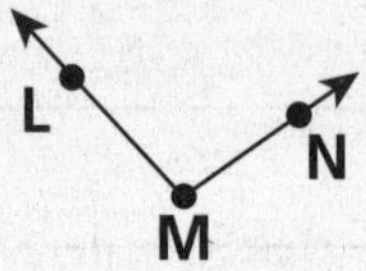

________ ________ ________

5.

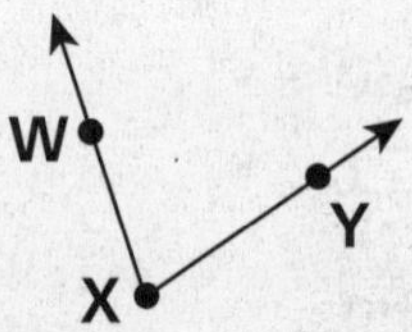

________ ________ ________

6.

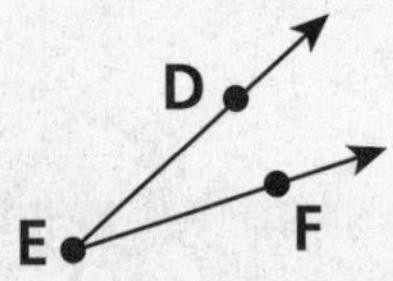

________ ________ ________

Rising Cubes

Directions **Read the information below. Then complete the page.**

Seneca works with a crew to build stage props in a local theater. They are using cubes to make risers for the cast to stand on. The front and right-side views are shown below.

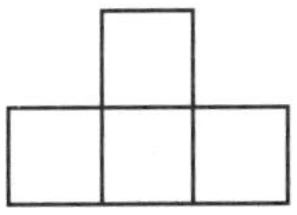

Front View

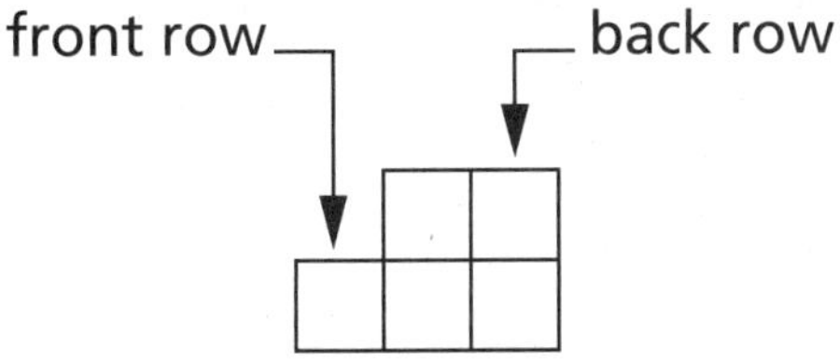

Right-side View

1. The figure is three rows deep from front to back.

a. What is the greatest number of cubes that can be in the front row? ________

b. What is the least number of cubes that can be in the front row? ________

2. On the grids below, draw the left-side view and the back view of the risers.

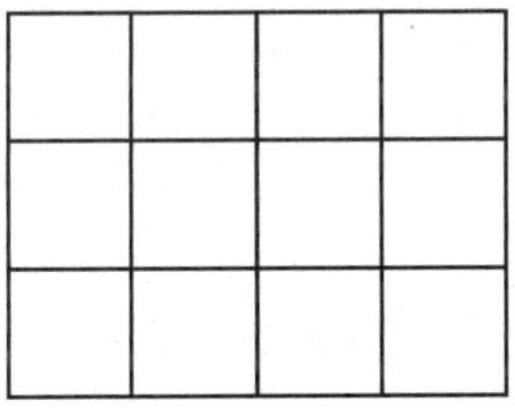

Left-side View

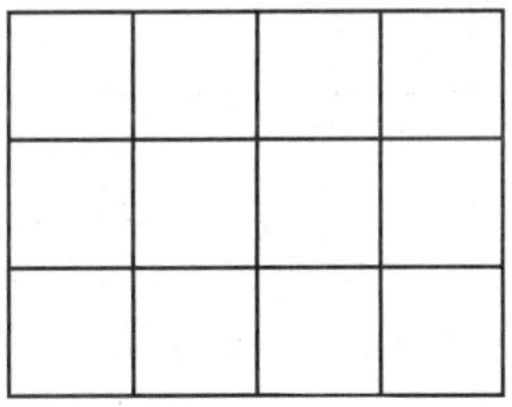

Back View

3. On the grids below, draw two possible top views of the figure.

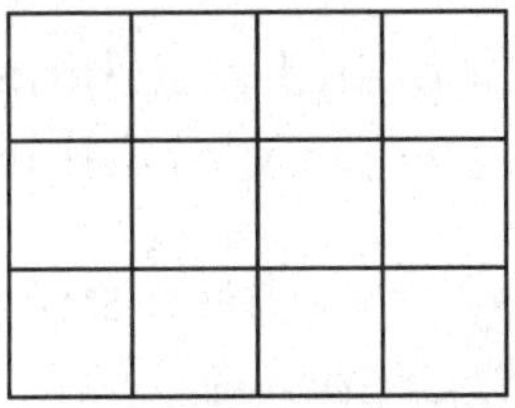

Possible Top View #1

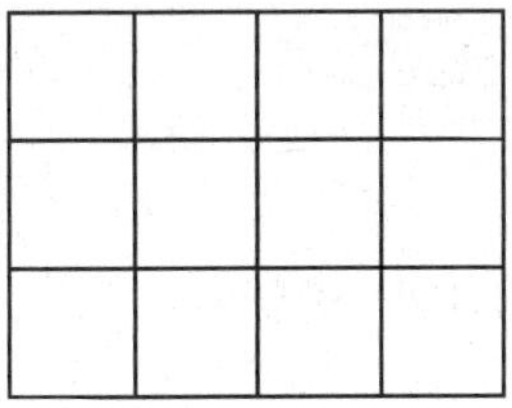

Possible Top View #2

Lights, Camera, Action!

Directions Read the clues. Use words from the box to complete the puzzle.

audition	script	rehearsal	crew	stage	recite
drama	comedy	audience	theater	applause	

ACROSS

1. People who build the props and work the lighting for plays
3. To try out for a part in a play or movie
4. A humorous play or television show
5. A practice run of a play prior to opening night
9. The place to watch a play or movie
10. Clapping to show approval

DOWN

2. To narrate or speak out loud, often from memory
3. If you are sitting and watching a play, you are a member of this
6. The screenplay or written story for a play, TV show, or movie
7. A raised platform where actors perform
8. A serious play or television show

Take a Bow

Olivia was waiting nervously to make her entrance onstage. As she peeked around the curtains, she saw that the community theater was filled. She knew that her family and friends were sitting in the audience, but there were also hundreds of strangers. "I shouldn't think about the audience," she reprimanded herself. "I should try to calm down."

Finally, the music began, and Olivia danced onstage. The only noise in the theater was the sound of Olivia's ballet shoes on the stage floor. Olivia felt her fear evaporate.

The audience was captivated by Olivia's performance. They marveled at the ballerina's grace and precision. However, Olivia was only conscious of the music and the movement of her body.

Finally, she heard the audience applauding and shouting, "Brava! Brava!" Olivia was elated as bouquets of flowers were placed at her feet. She took a deep bow.

Directions **Circle the letter of the best answer.**

1. How did Olivia feel before she went onstage?

A. scared **B.** calm

C. elated **D.** depressed

2. What is the meaning of the word **reprimanded** in the passage?

A. repeated **B.** argued

C. scolded **D.** questioned

3. Which of the following did **not** happen after Olivia started dancing?

A. Olivia performed well.

B. Olivia was nervous.

C. Olivia wasn't scared anymore.

D. The audience applauded.

4. Which of the following sentences in the passage is a compound sentence?

A. the first sentence **B.** the second sentence

C. the third sentence **D.** the fourth sentence

CHAPTER 10

Talk to Your Computer

Richard knew the Karachi Airport and the airspace around it well, having taken off from there several times before. His checklist was complete. He adjusted his headset, glanced at the cockpit dials, and looked down the runway.

"Tower, Cessna 916, Runway 14," he said. "Request takeoff clearance."

"Cleared for takeoff," came the reply in his earpiece. Richard opened the throttle.

Richard is playing a flight-simulation game that lets players experience the thrill of flying. It has two software improvements from the last version. One is that it stores and plays 25,000 pre-recorded audio clips, such as phrases air traffic controllers use when talking with pilots. These clips are spoken in the accented English of controllers from forty different parts of the world. The second is in the voice-recognition software. This software translates human speech into signals that a computer understands. This lets Richard feel like he's really controlling the plane.

Engineers who develop computer systems want computers to adapt to people, rather than people adapting to computers. This means they want the computers to communicate, or talk, in ways that seem natural to humans. These developers want people to be able to talk to their computers directly instead of using a device such as a keyboard or a mouse. These devices would be replaced by a human voice. However, getting computers to understand what humans say is very difficult.

To understand speech, a computer must recognize all the individual sounds that make up each word. For example, the spoken word *cat* has three sounds. It has a *k* sound at the beginning, a short *a* sound in the middle, and a *t* sound at the end. Many words that begin with the letter *c* have many different sounds within each word. To make it more difficult, people in different parts of the country, or the world, speak with different accents. English also sounds different when spoken in a continuous speech pattern instead of pronouncing each word, one-word-at-a-time.

Computers need a huge amount of processing power to sort out all of these sounds. Recently this power has become available at more affordable prices. This is why one of the first games to use voice-recognition software could only understand a few sets of phrases. With this game, it understands the phrases a pilot would use when talking to the control tower. If you read a poem or a grocery list to the game, it wouldn't know what to do!

Systems that understand several phrases are still very expensive. They require a much more powerful computer than many of us have. When those systems do become available, many people may still prefer to use the traditional keyboard and mouse. However, some people may like the freedom of using their voice instead of their hands. These voice-recognition systems can also offer independence to physically challenged people who cannot type on a keyboard or see the computer screen.

Voice-recognition software opens up a world of possibilities for many computer users. Imagine, "speaking" your next book report into your computer and having it type the words for you!

Talk to Your Computer

Directions Using what you have just read, answer the questions.

1. What is this article mostly about?

2. Why is it difficult to get computers to understand what people say?

3. Why were voice-recognition computers expensive in the past?

4. Why do you think only flight-simulator games are available?

5. Will voice-recognition software become more popular as the powerful computers become less expensive? Why or why not?

6. How might voice-recognition systems benefit physically challenged people?

Sentences

REMEMBER

A **simple sentence** has a complete subject and a complete predicate.
A **compound sentence** has two or more independent clauses.
A **complex sentence** has an independent clause and one or more dependent clauses.

EXAMPLES

Simple sentence: Modems and telephone lines let your computer "talk" to other computers.
Compound sentence: Jelani wants to learn about the Internet, but he doesn't know where to start.
Complex sentence: After you sign up with an online service, you'll get an e-mail address.

Sentence Connection

Directions **Read each sentence. Then write if it is a simple, compound, or complex sentence on the lines.**

1. An online service can connect you to the Internet. ____________
2. When you sign up with an online service, you have access to a browser. ____________
3. The World Wide Web is available to computers connected to the Internet. ____________
4. The Web is a collection of "pages." ____________
5. These pages have information and pictures, and they can also have sounds and motion. ____________
6. When you connect to a web site, the first page you see is called the home page. ____________
7. Links let you jump from one web page to another. ____________
8. Links look different from the rest of a page, and they are usually underlined and in a different color. ____________
9. You can try a link by clicking on it. ____________
10. "Surfing the Net" means cruising from one web page to another. ____________

Get Connected!

Directions **Choose words from the box below that match each definition. Write them on the lines.**

attachment	cautious	information	ingenious	instantly
opinion	remarkable	superhighway	worldwide	

1. Globally or around the world ____________________
2. Notable ____________________
3. A file enclosed within an e-mail message ____________________
4. Very creative and original ____________________

On the Internet

Directions **Read the article. Use words from the box above to fill in the blanks.**

Computers have changed our lives, from the way we do research for school to finding a time that a movie starts. We can find all kinds of **(5)** ____________________ at the click of a button. Millions of people get online to use the Internet. Almost **(6)** ____________________ they are transported all over the world. The Internet has become so popular that it is used by many people **(7)** ____________________, making it easy to communicate with someone on the other side of the globe. This connection has been called the information **(8)** ____________________, which means it's a roadway to learning. The Internet gives us easy access to historical occurrences, current events, and even games. However, because anyone can post information on the Internet, it is important to be **(9)** ____________________. This information may be someone's **(10)** ____________________ and not a fact!

Learning Capacity

REMEMBER

The main units of **capacity** in the U.S. Customary System are shown below.

U.S. Customary System

Units of Capacity	Abbreviation
1 fluid ounce	1 fl oz
1 cup (8 fl oz)	1 c
1 pint (2 c)	1 pt
1 quart (2 pt)	1 qt
1 gallon (4 qt)	1 gal

Measurement Match

Directions **Circle the containers that together will hold the same amount as the one in the box.**

1.

2.

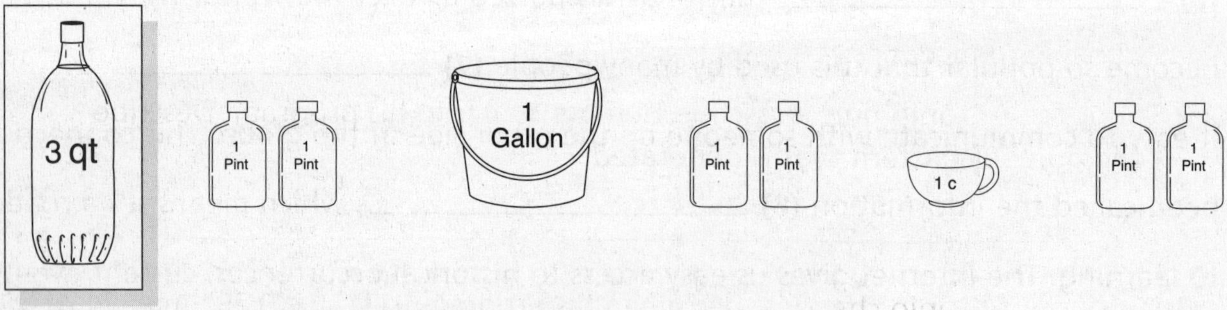

3.

Math Programs

Directions **Read the information below. Then complete the page.**

Antonio's computer program takes a number and changes it according to a rule. The rule now shown multiplies a number by 2 and then subtracts 4 from the product. The new number leaves through the "output" and is sent back to the starting point where it enters the machine again. This cycling continues until Antonio turns the program off. Each time Antonio starts the program with a new number, he lists the series of new numbers created as the program runs. Antonio likes to see the kind of number patterns that are created.

Computer Program

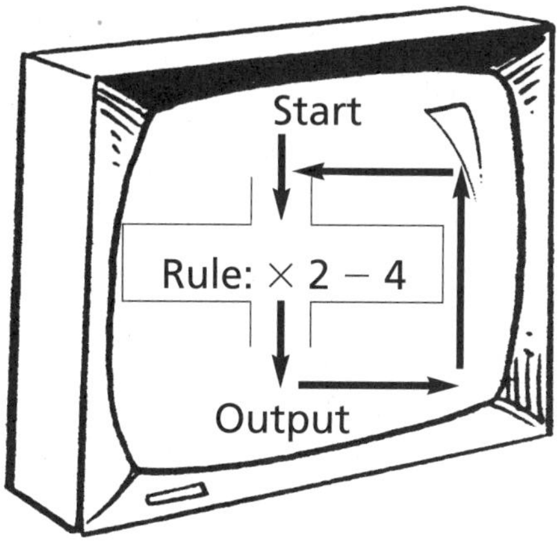

1. Suppose Antonio puts a starting number of 6 into his program.

a. Show how the first number leaving the program will be 8.

____

b. If 8 then enters the program, what number will be the next one to come out?

____

c. If Antonio's program continues to run, what will be the next four numbers that come out?

______, ______, ______, ______

2. Suppose Antonio puts a starting number of 4 into his program. Describe the number pattern that is generated.

____

3. Suppose Antonio changes the rule in his program to be "× 0.75" instead of "× 2 − 4." The rule also rounds each product to the nearest whole number.

a. Write the first 10 new numbers created using a starting number of 24.

____

b. What can you conclude about all new numbers after the tenth number?

____

Search Engine

Directions **Circle the words listed in the box. The words can be hidden across, down, or diagonally.**

attach	browser	computers	screen	Internet	link
online	modem	simulator	site	technology	web

l	a	t	e	c	h	n	o	l	o	g	y
m	i	r	l	u	d	b	i	a	r	s	i
x	s	n	a	z	o	l	n	v	e	c	n
r	w	s	k	s	a	p	t	w	p	r	m
r	e	i	o	c	r	o	e	h	o	e	o
c	o	m	p	u	t	e	r	s	v	e	d
d	s	u	o	e	a	o	n	l	i	n	e
d	m	l	t	s	i	t	e	r	a	b	m
r	b	a	r	n	r	z	t	c	w	n	e
w	n	t	y	p	e	c	z	a	i	e	w
d	a	o	a	g	y	z	e	r	c	l	b
s	b	r	o	w	s	e	r	w	p	h	t

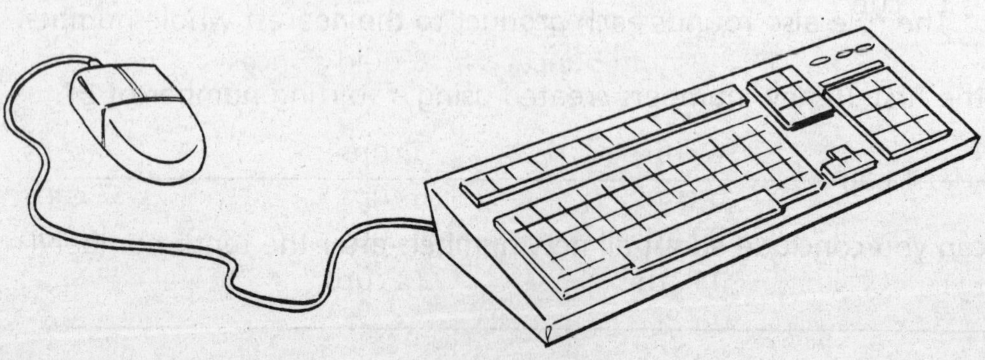

Capacity Combinations

Directions Circle the amounts that are equal to the quantity listed in the box. The first one has been done for you.

Example

	6 quarts	12 cups	3 pints	
		10 fluid ounces	1 quart	
		5 cups	16 fluid ounces	
1.	10 pints	2 quarts	4 quarts	6 quarts
		1 pint	3 pints	5 pints
		1 cup	2 cups	3 cups
2.	24 fluid ounces	5 fluid ounces	11 fluid ounces	3 cups
		15 fluid ounces	3 pints	
		1 cup	2 cups	
3.	5 cups	1 cup	2 cups	3 cups
		1 quart	2 quarts	3 quarts
4.	2 gallons	1 quart	2 quarts	14 fluid ounces
		5 quarts	8 fluid ounces	3 quarts
		10 fluid ounces	12 fluid ounces	
5.	2 gallons, 3 quarts	6 pints	7 pints	8 pints
		9 pints	19 pints	20 pints
		21 pints	22 pints	
6.	2 pints, 1 cup	1 pint	24 fluid ounces	16 fluid ounces
		4 fluid ounces	3 pints	5 pints
		12 fluid ounces	8 fluid ounces	
7.	2 quarts, 1 cup	2 cups	3 cups	4 cups
		5 cups	6 cups	7 cups
		8 cups	9 cups	10 cups
		11 cups	12 cups	

READING CHECK-UP

George Herman Ruth

On February 6, 1895, George Ruth was born in Baltimore, Maryland. His parents both died while he was young, so George was raised in an orphanage. It was there that he gained a reputation for his carefree ways and tough attitude.

One day, George happened to catch the eye of a movie producer. The producer was convinced that George would win the hearts of many fans. He promised to put George's name in the spotlight. George accepted an offer to star in the producer's next silent film. The movie was made in just a few weeks. George was now $25,000 richer, a hefty sum in those days.

George began to believe his own potential. He was so sure of himself, in fact, that he held onto the $25,000 check for several months to show it off to friends. Unfortunately, by the time George got around to cashing the check, the film had been released and was a complete failure. The producer's studio went bankrupt, and there was no money to cover the amount of the check. Though some people might have felt bitter, George was inspired. He decided to give up acting and follow his one true dream, to play professional baseball. In time, George did win the hearts of many fans and found a career that put his name in the spotlight. Today, George Ruth is best known by a slightly different name, "Babe" Ruth, one of the greatest baseball players of all time.

Directions **Fill in the bubble next to the correct answer.**

1. The author wrote this selection to tell how

Ⓐ silent movies were filmed.

Ⓑ people lived during the 1800s.

Ⓒ failure didn't stop one man from finding success.

Ⓓ a man dreamed of becoming a baseball player.

2. Another title for the passage is

Ⓐ "Pitcher Turned Movie Star"

Ⓑ "Carefree Youth"

Ⓒ "Great Football Players"

Ⓓ "Babe, Before Baseball"

3. George Ruth was born

Ⓐ during the twentieth century.

Ⓑ in Baltimore, Maryland.

Ⓒ in an orphanage.

Ⓓ to a family of actors.

4. Why did George keep the check he earned as an actor for so long?

Ⓐ He felt banks were unsafe.

Ⓑ He did not need the money.

Ⓒ He wanted to show it off.

Ⓓ He knew the studio had no money to cover the check.

READING CHECK-UP

The Assignment

"I can't believe it!" Shawna shouted as she opened the front door. She tossed her backpack onto the couch and threw her jacket onto the floor.

Shawna's grandmother rushed in from the kitchen to find out the cause of her granddaughter's frustration.

"What's the matter?" asked her grandmother.

"Mrs. Breen gave us another writing assignment. I've written so many now that my fingers are going to fall off! I have to interview a relative and write about an event that shaped that person's life. How boring!"

Shawna's grandmother just chuckled. "To be upset about such a small dilemma," she thought.

Shawna was hoping for some sympathy, but she received none from her grandmother.

"Trust me," said her grandmother, "you'll learn something valuable by interviewing a relative. You'll probably find out something you never even knew. Take your grandfather, for instance. He fought in the Vietnam War, he knew Martin Luther King, Jr., and he worked for President Carter."

"Really?" Shawna asked. She was surprised. She had never thought of her grandfather as a young person. "Gram, do you think I could interview Gramps for my assignment?"

"You should probably ask him," said her grandmother. "He's outside working in the garden."

"This is going to be great!" exclaimed Shawna as she grabbed her notebook and pen from her backpack. Then she raced out to the garden.

Directions **Fill in the bubble next to the correct answer.**

5. The word *dilemma* probably means

- Ⓐ assignment.
- Ⓑ predicament.
- Ⓒ complaint.
- Ⓓ distraction.

6. From the story, you can conclude that Shawna

- Ⓐ knew little about her grandfather's life.
- Ⓑ enjoyed these writing assignments.
- Ⓒ rarely speaks to her grandfather.
- Ⓓ always gets sympathy from her grandmother.

7. How does Shawna feel at the beginning of the story?

- Ⓐ enthusiastic
- Ⓑ alarmed
- Ⓒ frustrated
- Ⓓ overwhelmed

8. Shawna will most likely

- Ⓐ work with her grandfather in the garden.
- Ⓑ interview her grandfather.
- Ⓒ interview her grandmother.
- Ⓓ refuse to do the assignment.

READING CHECK-UP

Today's Customs

Many of our everyday customs can be traced back to ancient Greece. The tradition of lighting birthday candles was sparked by the Greek belief that placing lit candles on cakes helped ward off evil spirits. The idea of using perfume developed, in part, from ancient Greeks who rubbed spices on their bodies to give them a more pleasant smell in the hot climate. The custom of wearing wedding rings on the third finger also began in ancient Greece. Greeks believed that the third finger housed the "vein of love" which ran directly to the heart.

The ancient Greeks also popularized the idea of dyeing hair. They developed a way to dye their hair blonde to look like Greek gods and goddesses. The ancient Greeks also made early versions of toy dolls with moveable parts. They were also responsible for creating one of the most popular American desserts—pie! Of course, their pies differed slightly; they consisted of fish and meat. Finally, the word *marathon* traces back to a Greek soldier named Pheidippides (fI-dip-i-dEz). He ran 22 miles to spread the news of a Greek victory in the Battle at Marathon.

Directions **Fill in the bubble next to the correct answer.**

9. Which statement expresses an opinion?

Ⓐ Ancient Greeks made toy dolls.

Ⓑ Ancient Greeks were the most fascinating civilization.

Ⓒ Greek pies consisted of meat.

Ⓓ The word marathon traces back to ancient Greece.

10. Which of these Greek customs are no longer practiced today in America?

Ⓐ Candles are placed on birthday cakes

Ⓑ Wedding rings are worn on the third finger

Ⓒ Desserts are made with fish.

Ⓓ People run marathons

11. Why did ancient Greeks rub spices on their bodies?

Ⓐ Spices made them cool.

Ⓑ Spices helped them fall in love.

Ⓒ Spices helped them look like gods.

Ⓓ Spices gave their bodies a pleasant aroma.

12. The author wrote this passage to

Ⓐ describe how an ancient culture has influenced modern times.

Ⓑ convince people to learn more about ancient Greece.

Ⓒ inform people about Greece today.

Ⓓ explain the importance of customs in ancient Greece.

READING CHECK-UP
Suleiman the Magnificent

Before Columbus discovered the New World and Elizabeth I became the Queen of England, Suleiman the Magnificent ruled the Ottoman Empire. By 1520, this empire covered much of what is now the Middle East, northern Africa, Greece, eastern Europe, and southwestern Asia. By the end of Suleiman's first decade as Sultan, he had waged wars of expansion in Europe, Asia, and Africa. He controlled the major cities in northern Africa and, therefore, controlled the trade routes to the interior of the continent. The island of Rhodes and the entire eastern Mediterranean belonged to him. He also fought a fierce battle to capture Austria and Hungary. Although he lost, he was considered the most powerful monarch on earth by 1530.

Suleiman's huge empire required an efficient government and social organization. He created a complex military state designed to control the diverse cultures, religious groups, and territories that he had conquered. These territories were divided into millets, or small nations. The head of each millet reported to the Sultan and his staff. This system kept conflict among all the groups within each millet to a minimum. However, some believe it may have contributed to the ethnic tensions that exist in those areas today.

Suleiman ruled until 1566. He created harmony among the various groups under his rule, which was a significant achievement. Later in his reign, art, science, literature, and learning flourished.

Directions **Fill in the bubble next to the correct answer.**

13. What country didn't Suleiman the Magnificent invade?

Ⓐ Europe

Ⓑ Antarctica

Ⓒ Asia

Ⓓ Africa

14. From this selection, you can assume that Suleiman

Ⓐ was very intelligent.

Ⓑ liked art.

Ⓒ liked to fight.

Ⓓ was unfair.

15. According to the article, Suleiman the Magnificent was

Ⓐ rarely successful in battle.

Ⓑ a great religious leader and lawgiver.

Ⓒ the most powerful monarch.

Ⓓ the conqueror of Austria and Hungary.

16. Suleiman's huge empire required

Ⓐ new trade routes.

Ⓑ an efficient government.

Ⓒ a large castle.

Ⓓ several rulers.

Stop! Number Correct: ___ out of 16

MATH CHECK-UP

Directions **Read each question. Fill in the bubble next to the correct answer.**

1. $\begin{array}{r} 275{,}408 \\ -168{,}219 \\ \hline \end{array}$

(A) 13,211

(B) 7,189

(C) 17,189

(D) 107,189

2. $\begin{array}{r} 8{,}027 \\ \times 415 \\ \hline \end{array}$

(A) 3,331,205

(B) 3,230,105

(C) 3,330,905

(D) 3,303,950

3. $14\overline{)32{,}274}$

(A) 2,360

(B) 2,305 R4

(C) 235 R1

(D) 2,350 R4

4. $8\frac{5}{6} + 2\frac{1}{3} + 1\frac{1}{2} =$

(A) $4\frac{2}{9}$

(B) $25\frac{1}{3}$

(C) $12\frac{2}{3}$

(D) $11\frac{7}{11}$

5. $1\frac{4}{9} \div \frac{2}{3} =$

(A) $\frac{12}{3}$

(B) 2

(C) $\frac{26}{27}$

(D) $2\frac{1}{6}$

6. $29.806 + 8.27 + 143.1 =$

(A) 170.176

(B) 180.824

(C) 181.176

(D) 32.064

7. $\begin{array}{r} 0.182 \\ \times 2.7 \\ \hline \end{array}$

(A) 4.914

(B) 49.14

(C) 0.04914

(D) 0.4914

8. $47.8 + 2.645 + 127.09 =$

(A) 158.320

(B) 177.535

(C) 0.015832

(D) 178.325

9. There are 1,000 milligrams in a gram. There are 1,000 grams in a kilogram. How many milligrams are in 5 kilograms?

(A) 5,000

(B) 30,000

(C) 5

(D) 5,000,000

10. What is the area of the figure below?

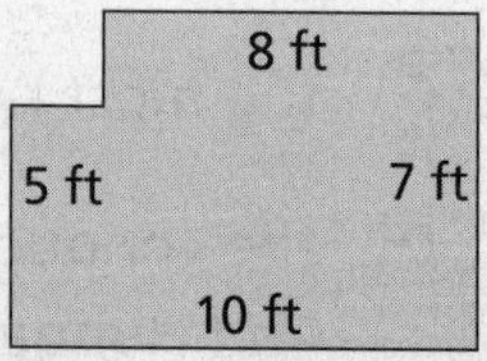

(A) 30 ft2

(B) 49 ft2

(C) 66 ft2

(D) 70 ft2

MATH CHECK-UP

Directions **Read each question. Fill in the bubble next to the correct answer.**

11. Carlos bought a pair of sneakers at a 15% discount. The sneakers usually cost $80. How much will they cost with the discount?

Ⓐ $92

Ⓑ $12

Ⓒ $82

Ⓓ $68

12. Name the pair of line segments that are perpendicular.

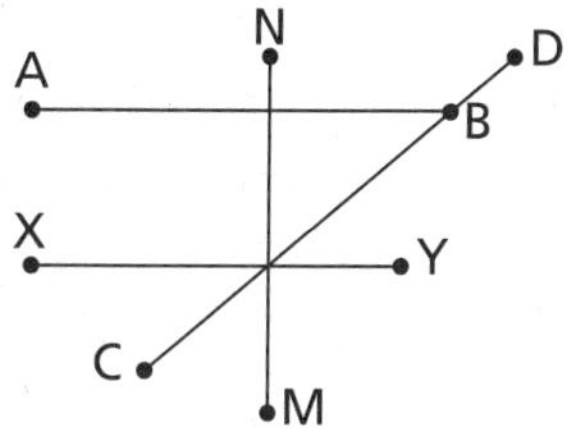

Ⓐ $\overline{AB}$ and $\overline{MN}$

Ⓑ $\overline{MN}$ and $\overline{CD}$

Ⓒ $\overline{XY}$ and $\overline{CD}$

Ⓓ $\overline{AB}$ and $\overline{XY}$

13. What is the measurement of $\angle P$?

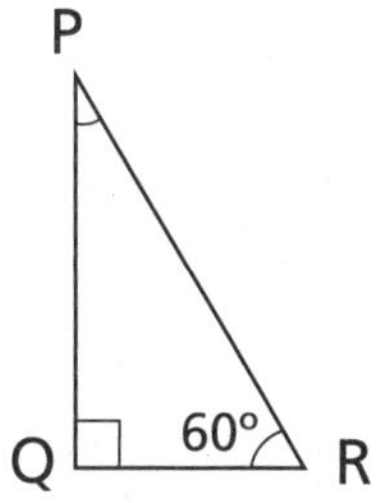

Ⓐ 12°

Ⓑ 30°

Ⓒ 45°

Ⓓ cannot be determined

14. If $m + n = 9$ and $m - n = n$, then

Ⓐ $m = 7, n = 2$

Ⓑ $m = 5, n = 4$

Ⓒ $m = 2, n = 27$

Ⓓ $m = 6, n = 3$

15. Diedre plans to attend sports camp. One week of camp costs $150. Children who attend 5 weeks pay $600 and get the 6th week free. What other information do you need to determine how much Diedre will pay?

Ⓐ the number of weeks Diedre will attend camp.

Ⓑ the number of weeks camp is in session.

Ⓒ the number of children attending camp.

Ⓓ the kind of sport Diedre plays.

16. Which point on the grid has the coordinates $(-5, -2)$?

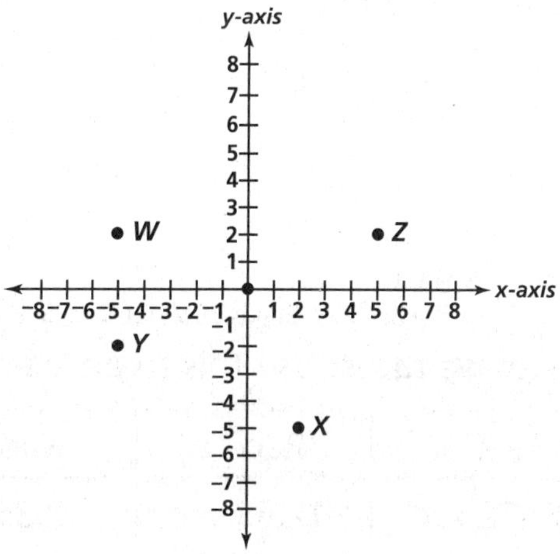

Ⓐ *W* Ⓑ *X* Ⓒ *Y* Ⓓ *Z*

17. $y^2 + y = 56$, so $y =$

Ⓐ 8

Ⓑ 7

Ⓒ 6

Ⓓ 5

18. All of the following are equal, **except**

Ⓐ $\frac{3}{4}$

Ⓑ 0.750

Ⓒ 75%

Ⓓ 750

MATH CHECK-UP

Directions **Read each question. Fill in the bubble next to the correct answer.**

The graph shows the average temperature in two cities during four consecutive months. Use the graph to answer questions 19 and 20.

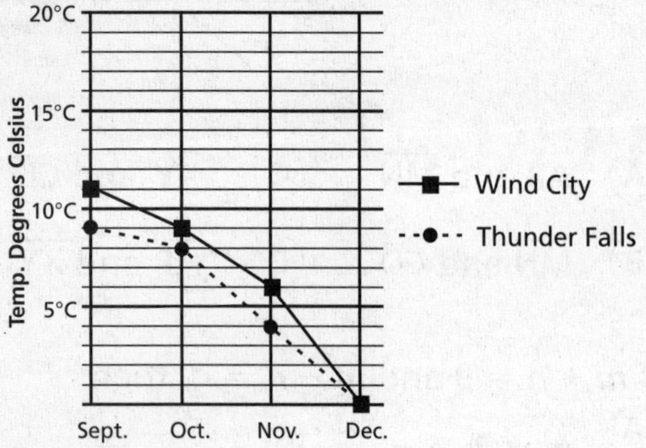

19 . How much warmer was Wind City than Thunder Falls during September?

Ⓐ 1°C

Ⓑ 2°C

Ⓒ 4°C

Ⓓ 5°C

20. Which two months in Wind City had a difference of 9°C?

Ⓐ Oct. and Dec.

Ⓑ Sept. and Dec.

Ⓒ Nov. and Dec.

Ⓓ Sept. and Nov.

21. Inside a bag there are twenty \$1 bills, seven \$5 bills, five \$10 bills, and three \$20 bills. If Jonah removes a bill from the bag without looking, what are his chances of removing a \$10 bill?

Ⓐ $\frac{1}{5}$

Ⓑ $\frac{5}{7}$

Ⓒ $\frac{1}{35}$

Ⓓ $\frac{1}{7}$

22. The chart below shows Ani's results for three races. Use this to answer

Race 1	Race 2	Race 3
12.52 sec	12.63 sec	12.35 sec

What is Ani's average race time?

Ⓐ 12.52 sec

Ⓑ 12.50 sec

Ⓒ 12.61 sec

Ⓓ 12.32 sec

23. On which point of the number line is the letter *y*?

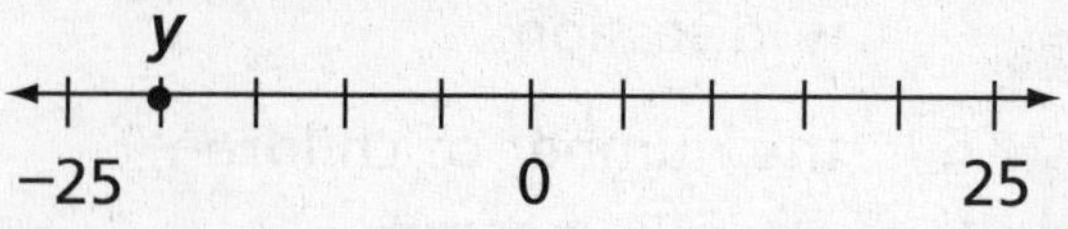

Ⓐ −20

Ⓑ −15

Ⓒ −10

Ⓓ 2

Stop! **Number Correct: ____ out of 23**

Congratulations!

____

(name)

has completed *Summer Counts!*

Good job!

Have a good school year.

ANSWER KEY

Page 5

1. He felt unlucky because he had to crawl on his stomach.
2. Buzzards never kill anything and clean up the countryside.
3. Possible answers: Eagle was jealous that Snake was flying. Eagle didn't like the change.
4. Snake grabbed for Dove.
5. Possible answer: Snake wanted to eat Dove.
6. Possible answers: The buzzards and Dove fly. They all were nice to Snake.
7. Answers will vary. Accept reasonable responses.

Page 6

1. accurately; comparative
2. motionless; comparative
3. most gracefully; superlative
4. clumsily; comparative
5. easily; comparative
6. more frequently; superlative
7. quickly; comparative
8. delicately; comparative
9. more ferociously; superlative
10. hungrily; comparative

Page 7

1. defenseless
2. aggressive
3. edible
4. prey
5. buoyant
6. predators
7. anatomy
8. aggressive
9. overpower
10. buoyant

Page 8

1. 1, 5; prime
2. 1, 2, 4, 8; composite
3. 1, 17; prime
4. 1, 2, 11, 22; composite
5. 1, 2, 3, 4, 6, 9, 12, 18, 36; composite
6. 1, 3, 17, 51; composite
7. 1, 73; prime
8. 1, 3, 9, 27, 81; composite
9. 1, 127; prime
10. 1, 3, 29, 87; composite

Page 9

1. 32.44
2. 34
3. 38
4. a. 14
   b. 9
   c. 2
5.

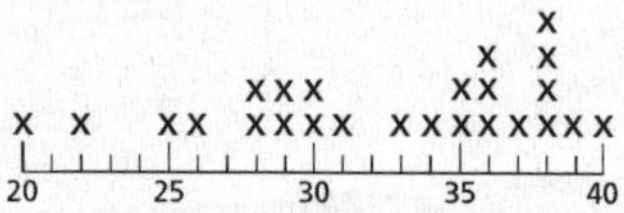

Page 10

1. beak week
2. nest quest
3. feather weather
4. hatch batch
5. pale tail
6. flight height

Page 11

1. 125
2. 117
3. 147
4. a. 3
   b. 7
   c. 7

Page 13

1. The brain controls the feelings.
2. Muscles contract and relax.
3. It goes to the arteries.
4. The heart has a left and a right side.
5. Both systems move blood. Pulmonary circulation pumps blood to the lungs; systemic circulation pumps blood to the rest of the body.
6. It increases the heart rate.
7. The author wrote this to tell about the heart.

Page 14

1. Recognizing
2. Looking at your choices
3. feeling
4. Accepting
5. Exercising
6. getting
7. communicating
8. Dealing with stress
9. Practicing
10. running

Page 15

1. prevent
2. calories
3. adolescence
4. confidence
5. encourage
6. healthy
7. diet
8. nutritious
9. calories
10. confidence

Page 16

1. $24 - 5 + 8 = 19 + 8 = 27$
2. $8 + 9 - 4 = 13$
3. $9 + 20 - 4 = 24$
4. $15 - 6 \times 8 \div 4 = 3$

Page 17

1. $\frac{2}{3}$
2.

E9	E10	E11
Susan	Lia	Monique
Susan	Monique	Lia
Lia	Susan	Monique
Lia	Monique	Susan
Monique	Susan	Lia
Monique	Lia	Susan

3. 6
4. $\frac{2}{3}$
5. $\frac{3}{100}$; any one of the girls can win.

Page 18

1. breath
2. headache
3. relax
4. homework
5. blood
6. stress
7. emotions
8. heart
9. plan

Riddle: Talk to someone!

Page 19

1. 3
2. 1
3. 3
4. 4
5. 4

Page 21

1. Rock climbing can be fun and exciting.
2. In traditional climbing, climbers place and remove gear. In sport climbing, they climb walls with fixed gear.
3. Possible answer: They won't fall far enough to get hurt.
4. Possible answer: Ropes need regular care so they're safe.
5. Answers will vary. Accept reasonable responses.
6. Possible answer: Brawn means strength.
7. The author thinks it is exciting and says this in the article.

Page 22

1. scorching
2. excited
3. scheduled
4. worn
5. aching
6. upcoming
7. waiting
8. parched
9. bottled
10. pounding

Page 23
1. dehydrated
2. energetic
3. agile
4. gear
5. endurance
6. muscles
7. circulation
8. jogger
9. surface
10. injury
11. dehydrated
12. agile

Page 24
1. no
2. yes
3. no
4. yes
5. 15
6. 4
7. 6
8. 35
9. 12
10. 44

Page 25
1. and 3.

	A	B
Sitting	0.2	1
Walking	0.6	4
Bicycling	1.2	5
Swimming	1.6	7
Jogging	2.0	11

2. 1.8
4. 4; 7

Page 26
1. soccer
2. rock climbing
3. track
4. high jumper
5. skiing
6. skateboarding
7. cheerleading

Page 27
1. C
2. B
3. B
4. D

Page 29
1. She was on a nursing assignment.
2. It's a collection of published letters.
3. She was more careful with them.
4. Possible answer: He probably read the newspaper column.
5. Possible answer: Yes, she wanted to stay there.
6. Possible answer: She will become a writer.
7. The author wrote this to tell about Nora Evangeline.

Page 30
1. to write
2. to read
3. to observe
4. To notice
5. to read
6. to improve
7. to attend
8. to help

Page 31
1. journalists
2. irresponsible
3. inaccurate
4. edit
5. translation
6. persuasive
7. viewpoint
8. convince
9. support
10. indifferent

Page 32
1. $0.\overline{333}$
2. 0.75
3. 0.625
4. $0.8\overline{333}$
5. $0.\overline{666}$
6. 0.8
7. $0.\overline{222}$
8. $0.58\overline{333}$

Page 33
1. a. length: 8.5
   width : 4.5
   b. length: 5.4
   width : 9.4
2. a. 26 inches
   b. 29.6 inches
3. a. 38.25 in.^2
   b. 50.76 in.^2

Page 34
1. stationary stationery
2. reed read
3. counsel council
4. fined find
5. tail tale
6. owed ode

Page 35
1. D
2. B
3. A
4. B

Page 37
1. Possible answer: It is about American political parties.
2. They are the Republicans and the Democrats.
3. Possible answer: They wanted to keep trade with England.
4. Possible answer: They were mostly farmers. They wanted to run their farms themselves.
5. Possible answers: Both are American political parties. Democrats oppose a strong central government. Republicans support a strong military.
6. Possible answer: Although there is no constitutional requirement for political parties, their rise is inevitable in a democracy.

Page 38
1. of designs; adjective
2. in geometric patterns; adverb
3. during the American Revolution; adverb
4. of the flag's stars and stripes; adjective
5. of our country; adjective
6. by many myths; adverb
7. of the first American flag; adjective
8. in a Washington, D.C. museum; adverb

Page 39
1. contributor
2. unlawful
3. veto
4. democracy
5. Constitution
6. envisioned
7. monarchy
8. government
9. vote
10. democracy

Page 40
1. (−4, 6)
2. (2, 2)
3. (0, −7)
4. (-5, −5)

5.–8.

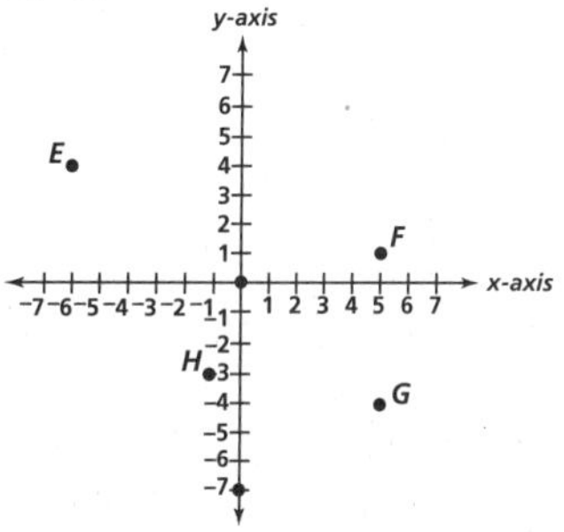

Page 41
1. a. $85.00
   b. $147.50
2. a.–b. Answers will vary. Accept reasonable responses.
3. Possible answer: Multiply each shirt sold by its cost. Then add.
4.

T-Shirts	Sweatshirts	Total Amount That Would be Collected
0	10	(0 x $8.50) + (10 x $14.75) = $147.50
1	9	(1 x $8.50) + (9 x $14.75) = $141.25
2	8	(2 x $8.50) + (8 x $14.75) = $135.00
3	7	(3 x $8.50) + (7 x $14.75) = $128.75
4	6	(4 x $8.50) + (6 x $14.75) = $122.50
5	5	(5 x $8.50) + (5 x $14.75) = $116.25
6	4	(6 x $8.50) + (4 x $14.75) = $110.00
7	3	(7 x $8.50) + (3 x $14.75) = $103.75
8	2	(8 x $8.50) + (2 x $14.75) = $97.50
9	1	(9 x $8.50) + (1 x $14.75) = $91.25
10	0	(10 x $8.50) + (0 x $14.75) = $85.00

5. a. 6
   b. 4

Page 42

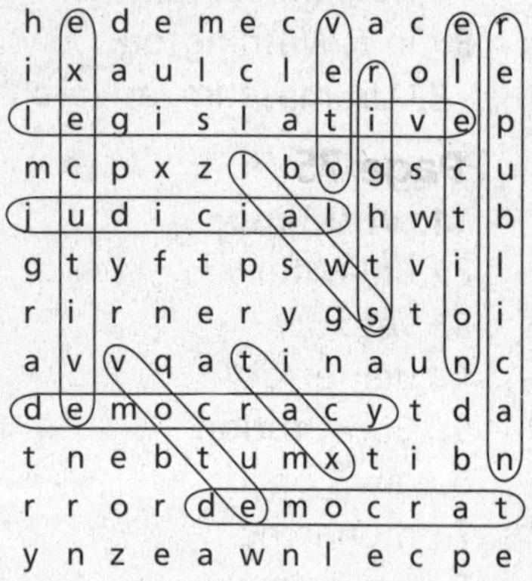

Page 43

1.–4.

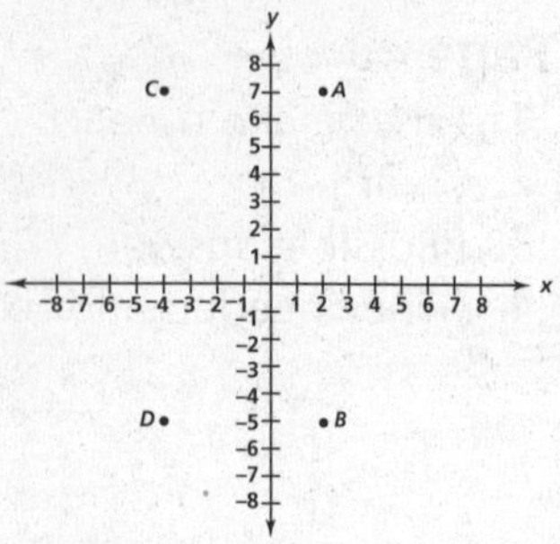

5. rectangle
6. 12 units
7. 6 units
8. 72 square units
9. Answers will vary. Accept reasonable responses.

Page 45

1. Mrs. Spencer hurt her back. She couldn't go to the store.
2. Min got her mail.
3. Possible answer: Min felt sorry for her. She wanted to help her.
4. Possible answer: Doing nice things for others will help you feel good about yourself.

5.–6. Answers will vary. Accept reasonable responses.

Page 46

1. independent
2. independent
3. independent
4. independent
5. dependent
6. independent
7. dependent

Page 47

1. consume
2. chef
3. aroma
4. whiff
5. originated
6. awakened
7. extinguished
8. custom
9. behaved
10. misbehaved

Page 48

1. 4 lines of symmetry
2. 2 lines of symmetry
3. 2 lines of symmetry
4. 0 lines of symmetry
5. 3 lines of symmetry
6. 8 lines of symmetry

Page 49

1. 28%
2. $\frac{28}{100} = \frac{7}{25}$
3. 0.28
4. 950; 190 ÷ 0.20
5. 10%
6. 38
7. 57

Page 50

1. great grate
2. doe dough
3. desert dessert
4. whole hole
5. Eight ate
6. Thyme time
7. Berry bury

Page 51

1. B
2. A
3. C
4. B

Page 53

1. It's about Susan B. Anthony's fight for African American and women's rights.
2. Possible answers: She felt the laws were unfair. She would not be any man's slave.
3. The Suffrage Movement was the first women's rights organization in the U.S.
4. Possible answer: Many people didn't want to change.
5. The Nineteenth Amendment gave women this right.
6. Possible answers: Women can vote; they have more rights; they can have any job they want.

Page 54

1. which was in a juice factory
2. who recognized his acting potential
3. which came as a surprise
4. which was Gibson's first movie role
5. whom he married in 1980
6. who seeks revenge for his father's murder
7. which is a film about a Scottish rebel
8. which earned awards for Best Director and Best Picture of 1995

Page 55

1. perspective
2. diligent
3. celebrity
4. biography
5. author
6. nonfiction
7. research
8. contemporary
9. challenging
10. libel

Page 56

1. a; d
2. b; d
3. a; b
4. b; c

Page 57

1.

Penny 1	Penny 2	Penny 3
heads	heads	heads
heads	heads	tails
heads	tails	heads
heads	tails	tails
tails	heads	heads
tails	heads	tails
tails	tails	heads
tails	tails	tails

2. 8
3. $\frac{1}{4}$; There are 8 combinations. Only 2 show the same side up.
4. three times
5. There are 3 combinations with 2 heads and 1 tail up. There's one where heads lands up 3 times.

Page 58

1. dancer
2. astronaut
3. actor
4. barber
5. scientist
6. singer
7. reporter
8. author
9. lawyer

Riddle: an autobiography

Page 59

1. 216
2. $\frac{1}{36}$
3. It is more likely to roll a different number each time. Answers will vary. Accept all reasonable responses.

Page 61

1. Possible answers: You have to learn what the customers want; how they want the shoes to make them feel.

2. Answers will vary. Accept reasonable responses.
3. Possible answer: A customer may like the way the shoes look, but they have to feel comfortable to buy them.

4.–5. Answers will vary. Accept reasonable responses.

6. Possible answer: Yes, he said he gets the most enjoyment out of this job.
7. Possible answer: The author wrote this to tell about the job of a shoe salesman.

Page 62
1. that can make cleaning easier; adjective clause
2. whomever is around; noun clause
3. that save time; adjective clause
4. Before you go to bed; adverb clause
5. what you don't use; noun clause
6. that comes last; adjective clause
7. what makes a room look messy; noun clause
8. whenever you're cleaning a room; adverb clause

Page 63
1. adversity
2. currency
3. wealthiest
4. bankrupt
5. tax
6. earnings
7. market
8. accounts
9. wealthiest
10. oversee

Page 64
1. yes
2. no
3. yes
4. no
5. yes

Page 65
1. $\frac{1}{6}$
2. $\frac{1}{8}$
3. $\frac{7}{24}$; $\frac{1}{6} + \frac{1}{8}$
4. Gerry and Keneisha work at the same pace.
5. about $3\frac{1}{2}$ hours

Page 66
1. cellar seller
2. tacks tax
3. bizarre bazaar
4. build billed
5. cash stash
6. fined find

Page 67
1. C
2. B
3. D
4. D

Page 69
1. They were sent to debtors' prison.
2. He went to school.
3. He sold prose sketches about life in London.
4. Possible answers: He did this through his writing. He ran charities.
5. Possible answers: He was poor when he was young. He had money as an adult.
6. Possible answer: An orator is a speaker.
7. Possible answers: He used humor to tell about the poor conditions in England. He wrote about common things.

Page 70
1. us
2. them
3. class
4. group
5. Mr. Brower
6. Marsha
7. us
8. Alyssa
9. us
10. students

Page 71
1. participate
2. converse
3. applause
4. audience
5. rhythm
6. poets
7. entirety
8. recite
9. stanzas
10. dramatic

Page 72
1. $\angle Y$, $\angle XYZ$, $\angle ZYX$
2. $\angle Q$, $\angle PQR$, $\angle RQP$
3. $\angle A$, $\angle CAT$, $\angle TAC$
4. $\angle M$, $\angle LMN$, $\angle NML$
5. $\angle X$, $\angle WXY$, $\angle YXW$
6. $\angle E$, $\angle DEF$, $\angle FED$

Page 73
1. **a.** 3
   b. 1
2.

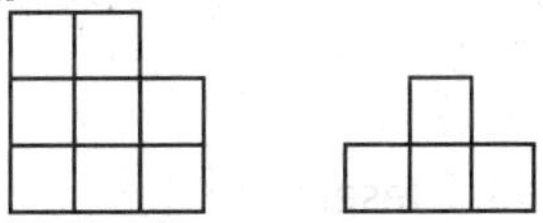

Left-side View   Back View

3. Possible answers

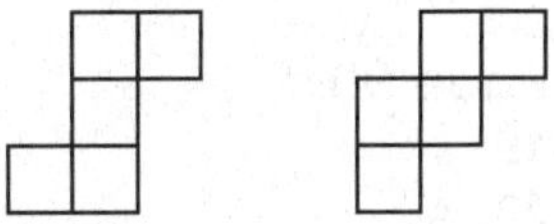

Possible Top View #1   Possible Top View #2

Page 74
Across
1. crew
3. audition
4. comedy
5. rehearsal
9. theater
10. applause

Down
2. recite
3. audience
6. script
7. stage
8. drama

Page 75
1. A
2. C
3. B
4. C

Page 77
1. Possible answer: It is about voice-recognition software.
2. Computers must recognize the individual sounds that make up each word.
3. Possible answer: They needed a lot of processing power.
4. Possible answer: Pilots already use this software for training.
5. Answers will vary. Accept reasonable responses.
6. Possible answer: People who can't use their hands can talk into the computer.

Page 78
1. simple
2. complex
3. simple
4. simple
5. compound
6. complex
7. simple
8. compound
9. simple
10. simple

Page 79
1. worldwide
2. remarkable
3. attachment
4. ingenious
5. information
6. instantly
7. worldwide

8. superhighway
9. cautious
10. opinion

Page 80

1. Circle 1c; 1c
2. Circle 2 pt; 2 pt; 2 pt
3. Circle 1qt; 2 pt; 1 qt; 2 pt

Page 81

1. a. 6 × 2 − 4 = 12 − 4 = 8
   b. 12; 8 × 2 − 4 = 16 − 4 = 12
   c. 20, 36, 68, 132
2. The pattern is a string of 4s.
3. a. 18, 14, 11, 8, 6, 5, 4, 3, 2, 2
   b. They will all be 2.

Page 82

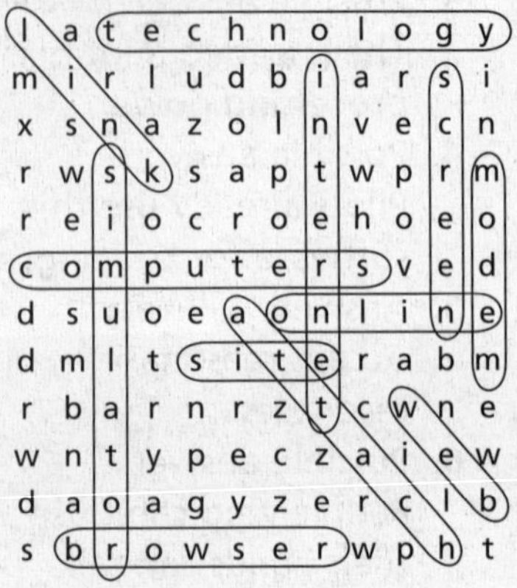

Page 83

1. 4 quarts, 1 pint, 2 cups; or 2 quarts, 1 pint, 5 pints
2. 1 cup, 2 cups; or 3 cups
3. 1 cup, 1 quart; or 2 cups, 3 cups
4. 1 quart, 3 quarts
5. 22 pints
6. 24 fluid ounces, 16 fluid ounces; or 1 pint, 16 fluid ounces, 8 fluid ounces
7. 9 cups

Page 84

1. C
2. D
3. B
4. C

Page 85

5. B
6. A
7. C
8. B

Page 86

9. B
10. C
11. D
12. A

Page 87

13. B
14. C
15. C
16. B

Page 88

1. D
2. A
3. B
4. C
5. D
6. C
7. D
8. B
9. D
10. D

Page 89

11. D
12. A
13. B
14. D
15. A
16. C
17. B
18. D

Page 90

19. B
20. A
21. D
22. B
23. A